Melbourne, Florida
www.paxen.com

Acknowledgements

For each of the selections and images listed below, grateful acknowledgement is made for permission to excerpt and/or reprint original or copyrighted material, as follows:

Text

58 From *The New York Times,* August 24, 2006 © The New York Times All rights reserved. Used by permission and protected by the Copyright Laws of the United States. The printing, copying, redistribution, or retransmission of the Material without express written permission is prohibited. **59** Used with the permission of the International Astronomical Union at www.iau.org/public_press/themes/pluto

Images

(cover, laptop) Kelly Redinger, DesignPics.com. **(cover, chameleon)** Corey Hochachka, DesignPics.com. **(cover, wind generators)** Don Hammond, DesignPics.com. **v** iStockphoto. **vi** iStockphoto. **x** Jamie Carroll/iStockphoto. **BLIND** Alex Wong/Getty Images. **42** Roger Ressmeyer/Corbis. **68** Steve C. Mitchell/Corbis.

ISBN-13: 978-1-934350-25-6
ISBN-10: 1-934350-25-7

2 3 4 5 6 7 8 9 10 GEDXSE1 16 15 14 13 12 11 10

Printed in the U.S.A.

Table of Contents

Title Page ... i
Copyright/Acknowledgements ... ii
Table of Contents ... iii
About the GED Tests ... iv–v
About *GED Prep Xcelerator* ... vi
About *GED Prep Xcelerator Science* ... vii
Test-Taking Tips ... viii
Study Skills ... ix
Before You Begin: Using Logic and Making Assumptions ... x–xi

UNIT 1 *Life Science*

GED Journeys: Richard H. Carmona ... 1

LESSON

- **1:** Interpret Tables ... 2–3
- **2:** Interpret Charts and Graphs ... 4–5
- **3:** Interpret Diagrams ... 6–7
- **4:** Interpret Illustrations ... 8–9
- **5:** Main Idea and Details ... 10–11
- **6:** Summarize ... 12–13
- **7:** Use Context Clues ... 14–15
- **8:** Analyze Information ... 16–17
- **9:** Categorize and Classify ... 18–19
- **10:** Compare and Contrast ... 20–21
- **11:** Sequence ... 22–23
- **12:** Interpret Timelines ... 24–25
- **13:** Cause and Effect ... 26–27
- **14:** Make Inferences ... 28–29
- **15:** Draw Conclusions ... 30–31
- **16:** Generalize ... 32–33
- **17:** Questioning ... 34–35

Unit 1 Review ... 36–41

UNIT 2 *Earth/Space Science*

GED Journeys: F. Story Musgrave ... 42–43

LESSON

- **1:** Understand and Evaluate a Hypothesis ... 44–45
- **2:** Interpret 3-Dimensional Diagrams ... 46–47
- **3:** Understand Maps and Map Symbols ... 48–49
- **4:** Interpret Physical and Topographic Maps ... 50–51
- **5:** Interpret Flowcharts ... 52–53
- **6:** Compare and Contrast Visuals ... 54–55
- **7:** Determine Fact and Opinion ... 56–57
- **8:** Evaluate Information ... 58–59
- **9:** Identify Problem and Solution ... 60–61

Unit 2 Review ... 62–67

UNIT 3 *Physical Science*

GED Journeys: Danica Patrick ... 68–69

LESSON

- **1:** Interpret Complex Diagrams ... 70–71
- **2:** Interpret Complex Tables ... 72–73
- **3:** Understand Illustrated Models ... 74–75
- **4:** Interpret Observations ... 76–77
- **5:** Predict Outcomes ... 78–79
- **6:** Use Calculations to Interpret Outcomes ... 80–81
- **7:** Draw Conclusions from Multiple Sources ... 82–83
- **8:** Interpret Multi-Bar and -Line Graphs ... 84–85
- **9:** Interpret Pictographs ... 86–87
- **10:** Relate Text and Figures ... 88–89
- **11:** Analyze Results ... 90–91
- **12:** Apply Concepts ... 92–93

Unit 3 Review ... 94–99

ANNOTATED ANSWER KEY ... 100–105

INDEX ... 106–110

About the GED Tests

Simply by turning to this page, you've made a decision that will change your life for the better. Each year, thousands of people just like you decide to pursue the General Educational Development (GED) certificate. Like you, they left school for one reason or another. And now, just like them, you've decided to continue your education by studying for and taking the GED Tests.

However, the GED Tests are no easy task. The tests—five in all, spread across the subject areas of Language Arts/Reading, Language Arts/Writing, Mathematics, Science, and Social Studies—cover slightly more than seven hours. Preparation takes considerably longer. The payoff, however, is significant: more and better career options, higher earnings, and the sense of achievement that comes with a GED certificate. Employers and colleges and universities accept the GED certificate as they would a high school diploma. On average, GED recipients earn $4,000 more per year than do employees without a GED certificate.

The American Council on Education (ACE) has constructed the GED Tests to mirror a high school curriculum. Although you will not need to know all of the information typically taught in high school, you will need to answer a variety of questions in specific subject areas. In Language Arts/Writing, you will need to write an essay on a topic of general knowledge.

In all cases, you will need to effectively read and follow directions, correctly interpret questions, and critically examine answer options. The table below details the five subject areas, the number of questions within each of them, and the time that you will have to answer them. Since different states have different requirements for the number of tests you may take in a single day, you will need to check with your local adult education center for requirements in your state or territory.

The original GED Tests were released in 1942 and have since been revised a total of three times. In each case, revisions to the tests have occurred as a result of educational findings or workplace needs. All told, more than 17 million people have received a GED certificate since the tests' inception.

SUBJECT AREA TEST	CONTENT AREAS	ITEMS	TIME LIMIT
Language Arts/Reading	Literary texts—75% Nonfiction texts—25%	40 questions	65 minutes
Language Arts/Writing (Editing)	Organization—15% Sentence Structure—30% Usage—30% Mechanics—25%	50 questions	75 minutes
Language Arts/Writing (Essay)	Essay	Essay	45 minutes
Mathematics	Number Sense/Operations—20% to 30% Data Measurement/Analysis—20% to 30% Algebra—20% to 30% Geometry—20% to 30%	Part I: 25 questions (with calculator) Part II: 25 questions (without calculator)	90 minutes
Science	Life Science—45% Earth/Space Science—20% Physical Science—35%	50 questions	80 minutes
Social Studies	Geography—15% U.S. History—25% World History—15% U.S. Government/Civics—25% Economics—20%	50 questions	70 minutes

Three of the subject-area tests—Language Arts/Reading, Science, and Social Studies—will require you to answer questions by interpreting passages. The Science and Social Studies tests also require you to interpret tables, charts, graphs, diagrams, timelines, political cartoons, and other visuals. In Language Arts/Reading, you also will need to answer questions based on workplace and consumer texts. The Mathematics Test will require you to use basic computation, analysis, and reasoning skills to solve a variety of word problems, many of them involving graphics. On all of the tests, questions will be multiple-choice with five answer options. An example follows:

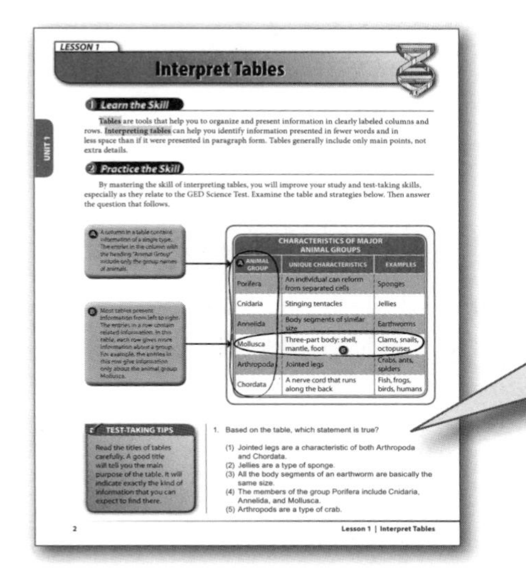

LESSON 1

Interpret Tables

UNIT 1

1 Learn the Skill

Tables are tools that help you to organize and present information in clearly labeled columns and rows. **Interpreting tables** can help you identify information presented in fewer words and in less space than if it were presented in paragraph form. Tables generally include only main points, not extra details.

2 Practice the Skill

By mastering the skill of interpreting tables, you will improve your study and test-taking skills, especially as they relate to the GED Science Test. Examine the table and strategies below. Then answer the question that follows.

CHARACTERISTICS OF MAJOR ANIMAL GROUPS

ANIMAL GROUP	UNIQUE CHARACTERISTICS	EXAMPLES
Porifera	An individual can reform from separated cells	Sponges
Cnidaria	Stinging tentacles	Jellies
Annelida	Body segments of similar size	Earthworms
Mollusca	Three-part body: shell, mantle, foot	Clams, snails, octopuses
Arthropoda	Jointed legs	Crabs, ants, spiders
Chordata	A nerve cord that runs along the back	Fish, frogs, birds, humans

TEST-TAKING TIPS

Read the titles of tables carefully. A good title will tell you the main purpose of the table. It will indicate exactly the kind of information that you can expect to find there.

1. Based on the table, which statement is true?

(1) Jointed legs are a characteristic of both Arthropoda and Chordata.
(2) Jellies are a type of sponge.
(3) All the body segments of an earthworm are basically the same size.
(4) The members of the group Porifera include Cnidaria, Annelida, and Mollusca.
(5) Arthropods are a type of crab.

2 Lesson 1 | Interpret Tables

Based on the table, which statement is true?

(1) Jointed legs are a characteristic of both Arthropoda and Chordata.
(2) Jellies are a type of sponge.
(3) All the body segments of an earthworm are basically the same size.
(4) The members of the group Porifera include Cnidaria, Annelida, and Mollusca.
(5) Arthropods are a type of crab.

On the Mathematics Test, you will have additional ways in which to register your responses to multiple-choice questions.

As the table on p. iv indicates, the Language Arts/Writing Test contains two parts, one for editing and the other for essay. In the editing portion of Language Arts/Writing, you will be asked to identify and correct common errors in various passages and texts while also deciding on the most effective organization of a text. In the essay portion, you will write an essay that provides an explanation or an opinion on a single topic of general knowledge.

So now that you understand the task at hand—and the benefits of a GED certificate—you must prepare for the GED Tests. In the pages that follow, you will find a recipe of sorts that, if followed, will help guide you toward successful completion of your GED certificate. So turn the page. The next chapter of your life begins right now.

About *GED Prep Xcelerator*

Along with choosing to pursue your GED certificate, you've made another smart decision by selecting *GED Prep Xcelerator* as your main study and preparation tool. Simply by purchasing *GED Prep Xcelerator*, you've joined an elite club with thousands of members, all with a common goal—earning their GED certificates. In this case, membership most definitely has its privileges.

For more than 65 years, the GED Tests have offered a second chance to people who need it most. To date, 17 million Americans like you have studied for and earned GED certificates and, in so doing, jump-started their lives and careers. Benefits abound for GED holders: Recent studies have shown that people with GED certificates earn more money, enjoy better health, and exhibit greater interest in and understanding of the world around them than do those without.

In addition, more than 60 percent of GED recipients plan to further their educations, which will provide them with more and better options. As if to underscore the point, U.S. Department of Labor projections show that 90 percent of the fastest growing jobs through 2014 will require postsecondary education.

Your pathway to the future—a *brighter* future—begins now, on this page, with *GED Prep Xcelerator*, an intense, accelerated approach to GED preparation. Unlike other programs, which take months to teach the GED Tests through a content-based approach, *Xcelerator* gets to the heart of the GED Tests—and quickly—by emphasizing *concepts*. At their core, the majority of the GED Tests are reading-comprehension exams. Students must be able to read and interpret excerpts, passages, and various visuals—tables, charts, graphs, timelines, and so on—and then answer questions based upon them.

Xcelerator shows you the way. By emphasizing key reading and thinking concepts, *Xcelerator* equips learners like you with the skills and strategies you'll need to correctly interpret and answer questions on the GED Tests. Two-page micro-lessons in each student book provide focused and efficient instruction, while callout boxes, sample exercises, and test-taking and other thinking strategies aid in understanding complex concepts. For those who require additional support, we offer the *Xcelerator* workbooks, which provide *twice* the support and practice exercises as the student books.

Unlike other GED materials, which were designed *for* the classroom, *Xcelerator* materials were designed *from* the classroom, using proven educational theory and cutting-edge classroom philosophy. The result: More than 90 percent of people who study with *Xcelerator* earn their GED certificates. For learners who have long had the deck stacked against them, the odds are finally in their favor. And yours.

GED BY THE NUMBERS

17 million
Number of GED recipients since the inception of GED Tests

1.23 million
Number of students who fail to graduate from high school each year

700,000
Number of GED test-takers each year

451,759
Total number of students who passed the GED Tests in 2007

$4,000
Additional earnings per year for GED recipients

About *GED Prep Xcelerator Science*

For those who think the GED Science Test is a breeze, think again. The GED Science Test is a rigorous exam that will assess your ability to understand and interpret subject-specific text or graphics. You will have a total of 80 minutes in which to answer 50 multiple-choice questions organized across three main content areas: Life Science (45% of all questions), Earth/Space Science (20%), and Physical Science (35%). Material in *GED Prep Xcelerator Science* has been organized with these percentages in mind.

GED Prep Xcelerator Science helps deconstruct the different elements of the test by helping learners like you build and develop key reading and thinking skills. A combination of targeted strategies, informational callouts and sample questions, tips and hints (Test-Taking Tips, Using Logic, and Making Assumptions), and ample assessment help to clearly focus study efforts in needed areas, all with an eye toward the end goal: success on the GED Tests.

As on the GED Social Studies Test, the GED Science Test uses the thinking skills of *comprehension*, *application*, *analysis*, and *evaluation*. In addition, items on the GED Science Test reflect the National Science Education Standards of Science as Inquiry; Science and Technology; Unifying Concepts and Processes; Science in Personal and Social Perspectives; and History and Nature of Science.

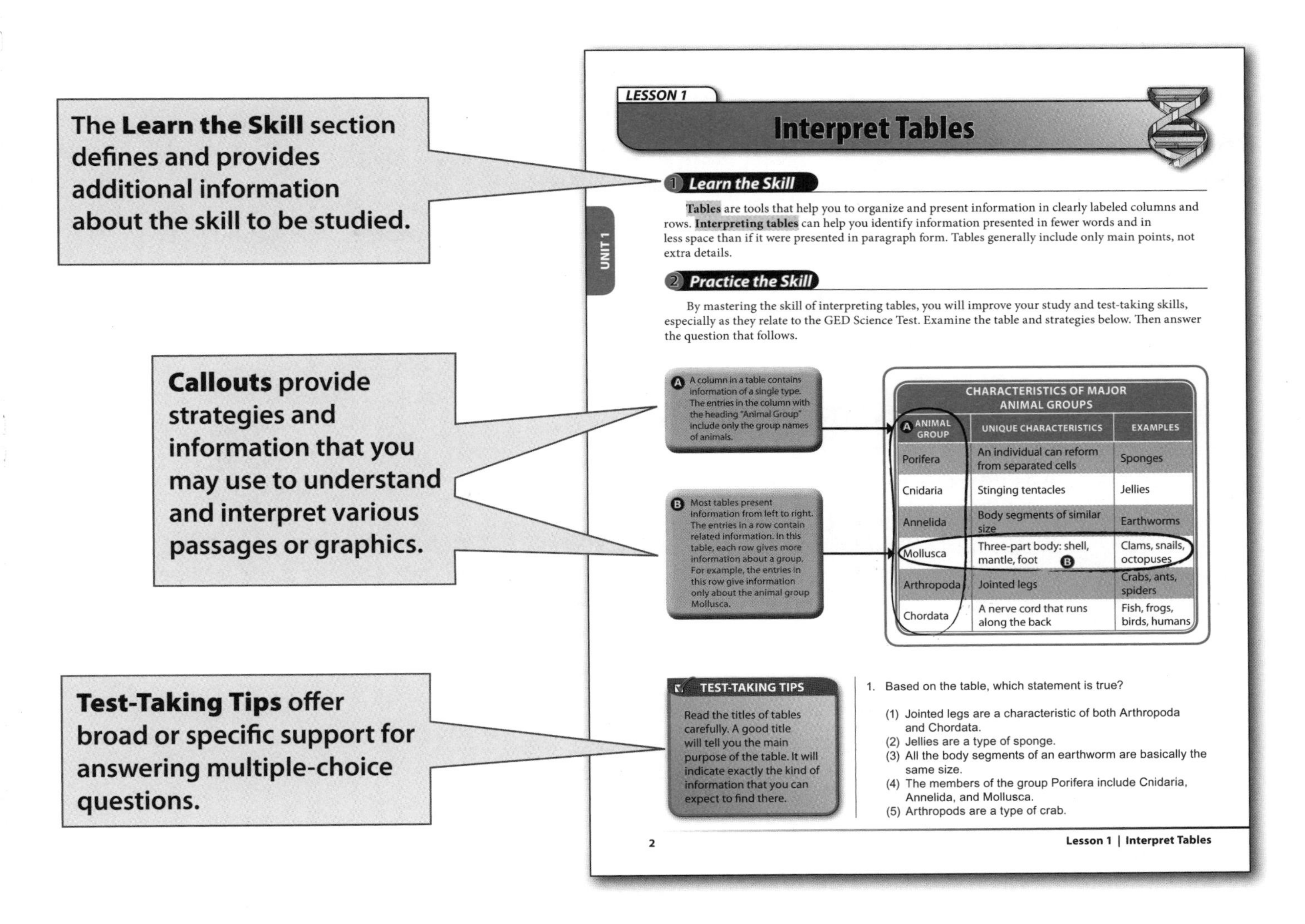

Test-Taking Tips

The GED Tests include 240 questions across the five subject-area exams of Language Arts/Reading, Language Arts/Writing, Mathematics, Science, and Social Studies. In each of the GED Tests, you will need to apply some amount of subject-area knowledge. However, because all of the questions are multiple-choice items largely based on text or visuals (such as tables, charts, or graphs), the emphasis in *GED Prep Xcelerator* is on helping learners like you build and develop core reading and thinking skills. As part of the overall strategy, various test-taking tips are included below and throughout the book to help you improve your performance on the GED Tests. For example:

- ***Always thoroughly read the directions so that you know exactly what to do.*** In Mathematics, for example, one part of the test allows for the use of a calculator. The other part does not. If you are unsure of what to do, ask the test provider if the directions can be explained.
- ***Read each question carefully so that you fully understand what it is asking.*** Some questions, for example, may present more information than you need to correctly answer them. Other questions may note emphasis through capitalized and boldfaced words (Which of the following is **NOT** an example of photosynthesis?).
- ***Manage your time with each question.*** Because the GED Tests are timed exams, you'll want to spend enough time with each question, but not *too* much time. For example, on the GED Science Test, you have 80 minutes in which to answer 50 multiple-choice questions. That works out to a little over 90 seconds per item. You can save time by first reading each question and its answer options before reading the passage or examining the graphic. Once you understand what the question is asking, review the passage or visual for the appropriate information.
- ***Note any unfamiliar words in questions.*** First, attempt to reread the question by omitting the unfamiliar word(s). Next, try to substitute another word in its place.
- ***Answer all questions, regardless of whether you know the answer or are guessing at it.*** There is no benefit in leaving questions unanswered on the GED Tests. Keep in mind the time that you have for each test and manage it accordingly. For time purposes, you may decide to initially skip questions. However, note them with a light mark beside the question and try to return to them before the end of the test.
- ***Narrow answer options by rereading each question and the text or graphic that goes with it.*** Although all five answers are *possible*, keep in mind that only one of them is *correct*. You may be able to eliminate one or two answers immediately; others may take more time and involve the use of either logic or assumptions. In some cases, you may need to make your best guess between two options. If so, keep in mind that test makers often avoid answer patterns; that is, if you know the previous answer is (2) and are unsure of the answer to the next question but have narrowed it to options (2) and (4), you may want to choose (4).
- ***Read all answer choices.*** Even though the first or second answer choice may appear to be correct, be sure to thoroughly read all five answer choices. Then go with your instinct when answering questions. For example, if your first instinct is to mark (1) in response to a question, it's best to stick with that answer unless you later determine that answer to be incorrect. Usually, the first answer you choose is the correct one.
- ***Correctly complete your answer sheet by marking one numbered space on the answer sheet beside the number that corresponds to it.*** Mark only one answer for each item; multiple answers will be scored as incorrect. If time permits, double-check your answer sheet after completing the test to ensure that you have made as many marks—no more, no less—as there are questions.

You've already made two very smart decisions in trying to earn your GED certificate and in purchasing *GED Prep Xcelerator* to help you to do so. The following are additional strategies to help you optimize your success on the GED Tests.

3 weeks out ...

- Set a study schedule for the GED Tests. Choose times in which you are most alert, and places, such as a library, that provide the best study environment.
- Thoroughly review all material in *GED Prep Xcelerator*, using the *GED Prep Xcelerator Science Workbook* to extend understanding of concepts in the *GED Prep Xcelerator Science Student Book*.
- Make sure that you have the necessary tools for the job: sharpened pencils, pens, paper, and, for Mathematics, the Casio FX-260 Solar calculator.
- Keep notebooks for each of the subject areas that you are studying. Folders with pockets are useful for storing loose papers.
- When taking notes, restate thoughts or ideas in your own words rather than copying them directly from a book. You can phrase these notes as complete sentences, as questions (with answers), or as fragments, provided you understand them.

1 week out ...

- Take the pretests, noting any troublesome subject areas. Focus your remaining study around those subject areas.
- Prepare the items you will need for the GED Tests: admission ticket (if necessary), acceptable form of identification, some sharpened No. 2 pencils (with erasers), a watch, eyeglasses (if necessary), a sweater or jacket, and a high-protein snack to eat during breaks.
- Map out the course to the test center, and visit it a day or two before your scheduled exam. If you drive, find a place to park at the center.
- Get a good night's sleep the night before the GED Tests. Studies have shown that learners with sufficient rest perform better in testing situations.

The day of ...

- Eat a hearty breakfast high in protein. As with the rest of your body, your brain needs ample energy to perform well.
- Arrive 30 minutes early to the testing center. This will allow sufficient time in the event of a change to a different testing classroom.
- Pack a sizeable lunch, especially if you plan to be at the testing center most of the day.
- Focus and relax. You've come this far, spending weeks preparing and studying for the GED Tests. It's your time to shine.

Before You Begin: Using Logic and Making Assumptions

At more than seven hours in length, the GED Tests are to testing what marathons are to running. Just like marathons, though, you may train for success on the GED Tests. As you know, the exams test your ability to interpret and answer questions about various passages and visual elements. Your ability to answer such questions involves the development and use of core reading and thinking skills. Chief among these are the skills of reasoning, logic, and assumptions.

Reasoning involves the ability to explain and describe ideas. **Logic** is the science of correct reasoning. Together, reasoning and logic guide our ability to make and understand assumptions. An **assumption** is a belief that we know to be true and which we use to understand the world around us.

You use logic and make assumptions every day, sometimes without even knowing that you're doing so. For example, you might go to bed one night knowing that your car outside is dry; you might awaken the next morning to discover that your car is wet. In that example, it would be *reasonable* for you to *assume* that your car is wet because it rained overnight. Even though you did not see it rain, it is the most *logical* explanation for the change in the car's appearance.

When thinking logically about items on the GED Tests, you identify the consequences, or answers, from text or visuals. Next, you determine whether the text or visuals logically and correctly support the consequences. If so, they are considered valid. If not, they are considered invalid. For example, read the following text and determine whether it is valid or invalid:

Passage A

The GED Tests assess a person's reading comprehension skills. Ellen enjoys reading. Therefore, Ellen will do well on the GED Tests.

Passage B

The GED Tests cover material in five different subject areas. Aaron has geared his studies toward the tests, and he has done well on practice tests. Therefore, Aaron may do well on the GED Tests.

Each of the above situations has a consequence: *Ellen will* or *Aaron may* do well on the GED Tests. By using reasoning and logic, you can make an assumption about which consequence is valid. In the example above, it is *un*reasonable to assume that Ellen will do well on the GED Tests simply because she likes to read. However, it *is* reasonable to assume that Aaron may do well on the GED Tests because he has studied for the tests and has done well on the practice tests in each of the five subject areas.

Use the same basic principles of reasoning, logic, and assumptions to determine which answer option logically and correctly supports a question on the GED Science Test. You may find occasions in which you have narrowed the field of possible correct answers to two, from which you must make a best, educated guess. In such cases, weigh both options and determine the one that, reasonably, makes the most sense.

Directions: Following are a series of questions to use in practicing the skills of reasoning, logic, and assumptions. The first one has been done for you.

CHARACTERISTICS OF MAJOR ANIMAL GROUPS		
ANIMAL GROUP	UNIQUE CHARACTERISTICS	EXAMPLES
Porifera	An individual can reform from separated cells	Sponges
Cnidaria	Stinging tentacles	Jellies
Annelida	Body segments of similar size	Earthworms
Mollusca	Three-part body: shell, mantle, foot	Clams, snails, octopuses
Arthropoda	Jointed legs	Crabs, ants, spiders
Chordata	A nerve cord that runs along the back	Fish, frogs, birds, humans

A You can eliminate some answer options, such as (1) and (5), simply by thoroughly reading the table.

B In some cases, answer options use more general statements. You must review the information, consider it logically, make comparisons, and construct an assumption from it. In the case of option (3), even though the table does not directly prove or disprove the answer choice, you must use the information here and your own knowledge to assume that other forms of life are more sophisticated.

C Some answer options require you to use logic and reasoning, such as (2) and (4). You must determine whether these statements are logical.

1. Which conclusion can you come to based only on the information in the table?

   (1) Because scorpions have jointed legs, they belong to the group Porifera. **A**
   (2) A dog has four feet, so it belongs to group Mollusca.
   (3) Animals that can reform from single cells are the most sophisticated forms of life. **B**
   (4) Sea anemones have stinging tentacles, so they belong to group Cnidaria.
   (5) Animals in Mollusca, Arthropoda, and Chordata all have at least one foot. **A**

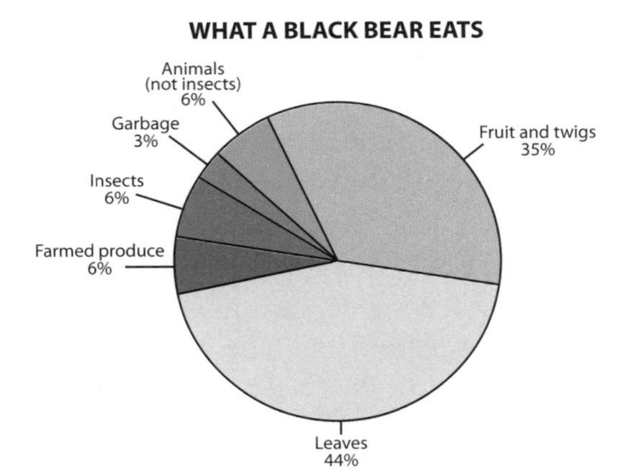

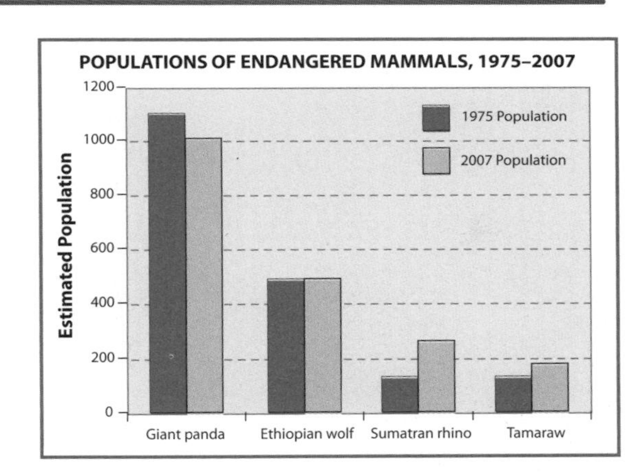

2. Which of the following statements is best supported by the data above?

   (1) Black bears often eat mice, chickens, and rabbits.
   (2) Berries are a black bear's favorite food.
   (3) Black bears might have a hard time finding food in the winter.
   (4) No other animals eat black bears.
   (5) Black bears can survive on fruit and twigs alone.

3. What can you conclude based on the graph above?

   (1) There were about as many Ethiopian wolves in 2007 as in 1975.
   (2) The tamaraw is prey for the Ethiopian wolf.
   (3) The giant panda population will continue to decrease each year.
   (4) The giant panda should no longer be listed as endangered.
   (5) There were more Sumatran rhinos than Giant pandas in 2007.

Answers: **2.** (3); **3.** (1)

GED JOURNEYS

RICHARD H. CARMONA

Richard Carmona used his GED certificate as a springboad to many career successes, including his service as the U.S. Surgeon General.

It takes about an hour to travel by air from New York City to Washington, D.C. For Richard Carmona, however, the journey to our nation's capital took considerably longer. Carmona, the U.S. Surgeon General from 2002 to 2006, went from one of our country's toughest neighborhoods to one of its highest positions.

As a teenager in Harlem, New York, Carmona left school without his high school diploma. At age 17, he enlisted in the U.S. Army. While in the military, Carmona earned his GED credential and joined the Special Forces, winning two Purple Hearts during the Vietnam War.

After leaving active duty, he earned an associate's degree from the Bronx (N.Y.) Community College. Carmona later received his undergraduate degree (1977) and M.D. (1979) from the University of California at San Francisco, where he also was named that school's top graduate. Carmona also earned a master's degree in public health from the University of Arizona. Carmona explained his desire to achieve:

> **"It's my sense of having to make up for lost time."**

In 1985, Carmona moved to Arizona, where he started the region's first trauma care program. He also worked as a paramedic, a registered nurse, a physician, and a SWAT team member. Carmona's extensive experience, including expertise in terrorist and biological threats, led to his appointment as the 17th Surgeon General of the U.S. Public Health Service in August 2002. In his four years as U.S. Surgeon General, Carmona served as our country's chief health educator. Among other achievements, Carmona worked to inform Americans about the need to prepare for natural and other disasters.

BIO BLAST: Richard Carmona

- Born in 1949 in Harlem, New York
- Became first member of his family to earn a college degree
- Served as Director of Trauma Services at Tucson Medical Center
- Worked as a doctor and SWAT team leader for the Pima County Sheriff's Department
- Honored in 2000 as a "top cop" by the National Association of Police Organizations
- Taught as a clinical professor at the University of Arizona in Tucson
- Served as a member of the Arizona Board of Medical Examiners
- Served as U.S. Surgeon General from 2002 to 2006

Unit 1: Life Science

Whenever you eat lunch, go to the gym, or even take a nap, you are using pieces of information from life science to guide and enrich your well-being. On a much larger scale, life science enables us to understand ourselves and our environment.

Similarly, life science plays an important part in the GED Science Test, comprising 45 percent of all questions. As with other areas of the GED Science Test, life science questions will test your ability to interpret text or graphics and to answer questions about them by using thinking skills such as comprehension, application, analysis, and evaluation. In Unit 1, the introduction of core reading skills and graphics, combined with essential science concepts, will help you prepare for the GED Science Test.

Table of Contents

Lesson 1: Interpret Tables 2–3
Lesson 2: Interpret Charts and Graphs 4–5
Lesson 3: Interpret Diagrams 6–7
Lesson 4: Interpret Illustrations 8–9
Lesson 5: Main Idea and Details 10–11
Lesson 6: Summarize 12–13
Lesson 7: Use Context Clues 14–15
Lesson 8: Analyze Information 16–17
Lesson 9: Categorize and Classify 18–19
Lesson 10: Compare and Contrast 20–21
Lesson 11: Sequence 22–23
Lesson 12: Interpret Timelines 24–25
Lesson 13: Cause and Effect 26–27
Lesson 14: Make Inferences 28–29
Lesson 15: Draw Conclusions 30–31
Lesson 16: Generalize 32–33
Lesson 17: Questioning 34–35
Unit 1 Review 36–41

LESSON 1

Interpret Tables

1 Learn the Skill

Tables are tools that help you to organize and present information in clearly labeled columns and rows. **Interpreting tables** can help you identify information presented in fewer words and in less space than if it were presented in paragraph form. Tables generally include only main points, not extra details.

2 Practice the Skill

By mastering the skill of interpreting tables, you will improve your study and test-taking skills, especially as they relate to the GED Science Test. Examine the table and strategies below. Then answer the question that follows.

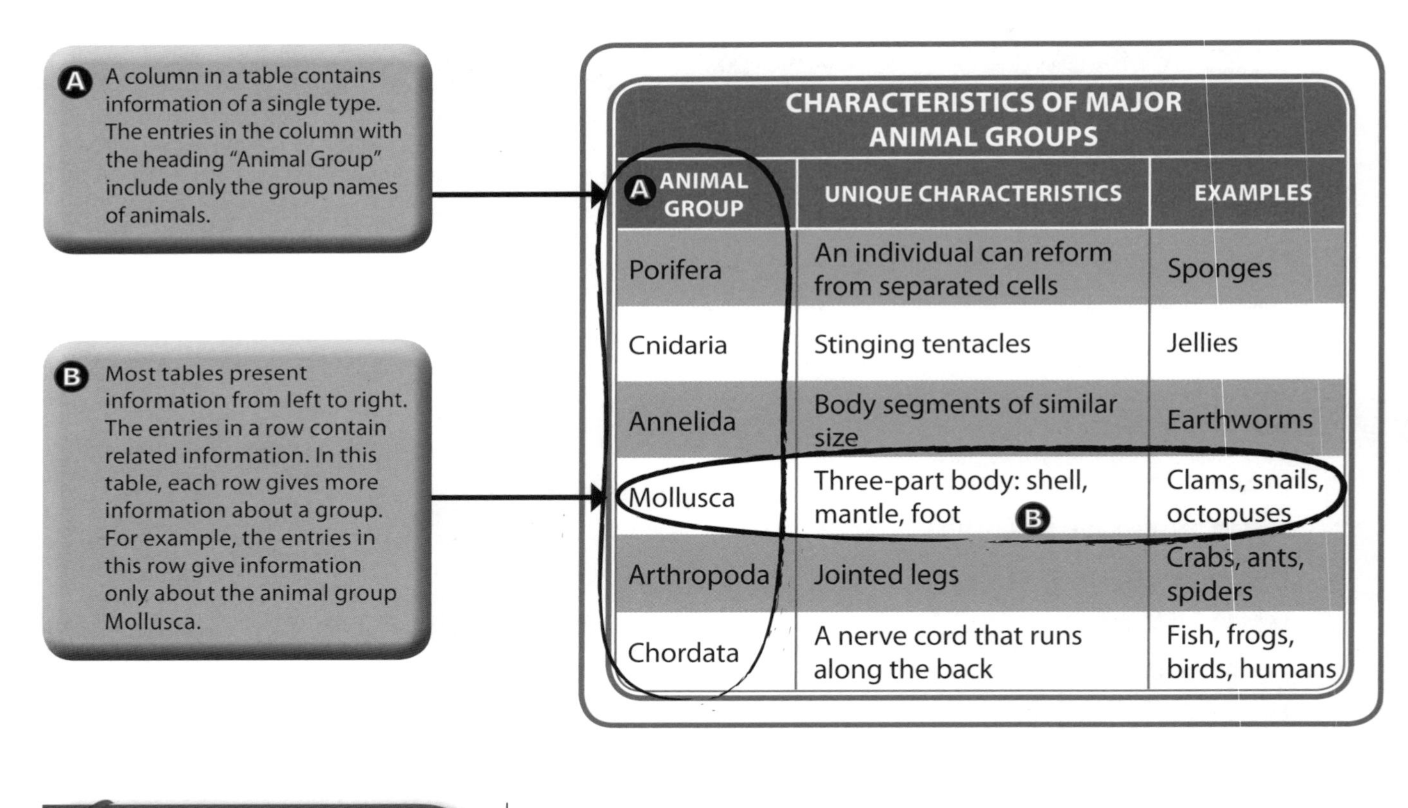

CHARACTERISTICS OF MAJOR ANIMAL GROUPS

ANIMAL GROUP	UNIQUE CHARACTERISTICS	EXAMPLES
Porifera	An individual can reform from separated cells	Sponges
Cnidaria	Stinging tentacles	Jellies
Annelida	Body segments of similar size	Earthworms
Mollusca	Three-part body: shell, mantle, foot	Clams, snails, octopuses
Arthropoda	Jointed legs	Crabs, ants, spiders
Chordata	A nerve cord that runs along the back	Fish, frogs, birds, humans

TEST-TAKING TIPS

Read the titles of tables carefully. A good title will tell you the main purpose of the table. It will indicate exactly the kind of information that you can expect to find there.

1. Based on the table, which statement is true?

   (1) Jointed legs are a characteristic of both Arthropoda and Chordata.
   (2) Jellies are a type of sponge.
   (3) All the body segments of an earthworm are basically the same size.
   (4) The members of the group Porifera include Cnidaria, Annelida, and Mollusca.
   (5) Arthropods are a type of crab.

3 Apply the Skill

Directions: Choose the one best answer to each question.

Questions 2 through 6 refer to the following information and table.

Chordates are animals with nerve cords that run along their backs. Most chordates have backbones that protect the nerve cords and provide the body with support. The main groups of chordates are fish, amphibians, reptiles, birds, and mammals.

Fish generally spend their entire lives in water. Amphibians generally begin life in water but spend their adult lives on land. Most reptiles spend their entire lives on land. Some groups of birds and mammals live in water, but the majority live on land.

GROUP	BODY COVER	BREATHING STRUCTURES	REPRODUCTION
Fish	Scales	Gills	Some lay eggs; some give birth to live young
Amphibians	Thin skin	Gills early in life; lungs	Lay eggs with no shells
Reptiles	Scaly skin	Lungs	Lay eggs with leathery shells
Birds	Feathers	Lungs	Lay eggs with brittle shells
Mammals	Hair or fur	Lungs	Give birth to live young (most)

2. Which of the following is the best title for the table?

    (1) Examples of Chordates
    (2) Animals with Backbones
    (3) Characteristics of Animals
    (4) Characteristics of Chordate Groups
    (5) How Major Animal Groups Differ

3. Which of the following would be an appropriate heading for a fifth column of the table?

    (1) Non-Chordate Groups
    (2) Insects
    (3) Fur Color
    (4) Kind of Eggs They Lay
    (5) Types of Limbs

4. Based on the table, which animal groups are most alike?

    (1) fish and mammals
    (2) amphibians and fish
    (3) birds and fish
    (4) mammals and birds
    (5) reptiles and birds

5. Based on the information and table, which statement is true?

    (1) Chordates are the only animals that lay eggs.
    (2) Fish are the only chordates that live in water.
    (3) Hair is a defining characteristic of chordates.
    (4) The majority of chordates rely on lungs.
    (5) All chordates have scales, thin skin, feathers, and hair.

6. Which statement describes the best way to change the table to include all the information in the paragraph?

    (1) Add a new row to the bottom of the table entitled "Animals that Live in Water."
    (2) Add more detailed information to the column "Breathing Structures."
    (3) Add a column with the heading "Main Habitat."
    (4) Replace the last column of the table with a column called "Structures for Movement."
    (5) Replace the column heading "Body Cover" with the heading "Water Preference."

Interpret Charts and Graphs

1 Learn the Skill

Charts and graphs are visual ways to show data. **Interpreting charts and graphs** can help you answer questions about data. Unlike tables, which are generally used to show data as text, graphs and charts visually depict parts of a whole or changes over time.

2 Practice the Skill

By mastering the skill of interpreting charts and graphs, you will improve your study and test-taking skills, especially as they relate to the GED Science Test. Study the paragraph and circle graph below. Then answer the question that follows.

A Circle graphs show parts of a whole. The sum of the percentages of the sections in a circle graph is always 100%. Sometimes, the sections are color-coded and correspond to a legend or key. In this example, the sections are individually labeled.

B The relative sizes of the sections of a circle graph provide visual clues about the relationships among the sections. In this graph, the largest sections represent the foods that black bears eat most often. The smallest sections represent the foods that black bears eat least often.

Black bears live in many parts of the United States. Many people think that black bears eat mainly meat and berries. However, their diets can vary greatly. The circle graph shows the different kinds of foods a typical black bear may eat.

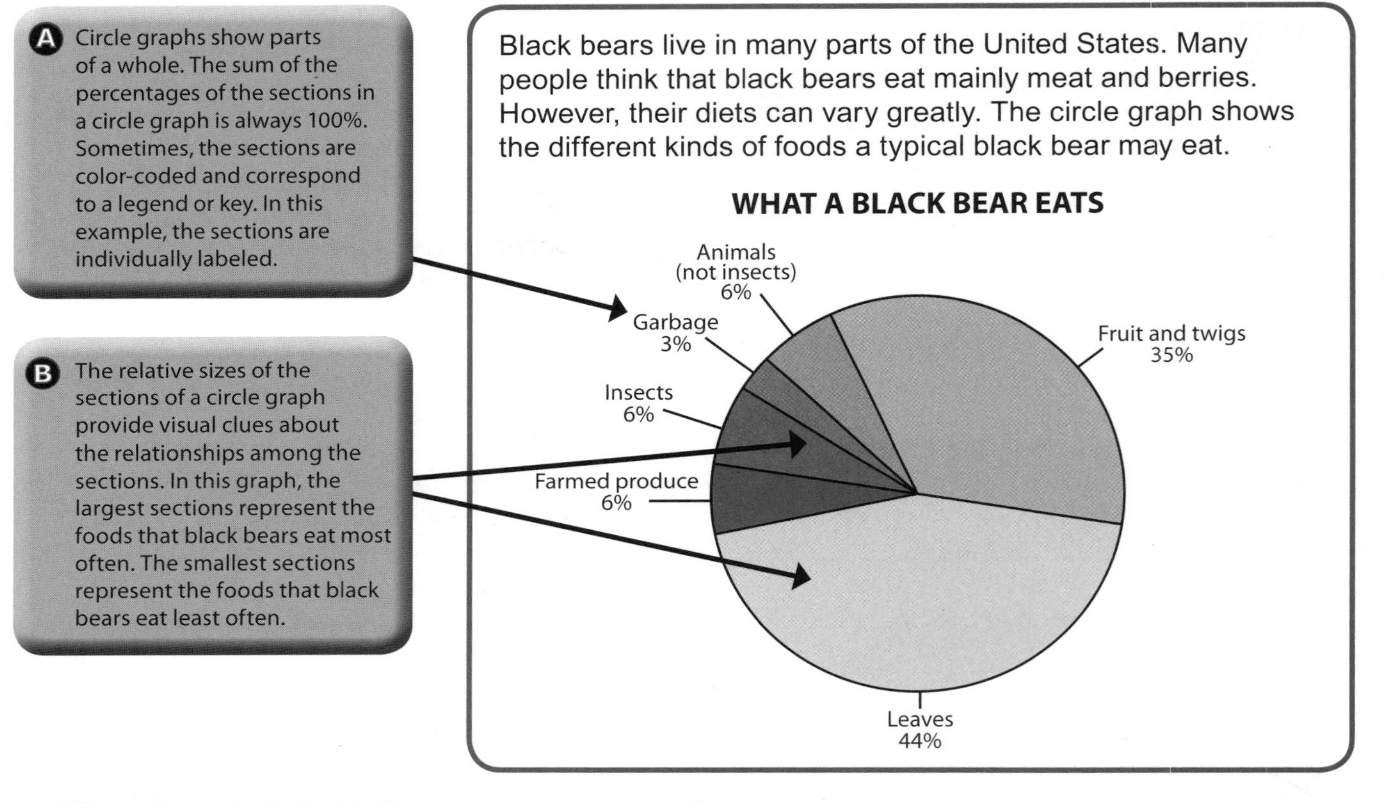

TEST-TAKING TIPS

When interpreting a graph, always read the title and labels on the graph carefully. The title and labels give important information about the data in the graph. If the graph has a legend, or key, study it carefully, too.

1. Based on the information, which of the following statements about a black bear's diet is true?

   A black bear's diet

   (1) contains more meat than plants
   (2) is mainly insects and animals
   (3) consists of equal amounts of garbage and insects
   (4) consists mainly of fruit and other plants
   (5) contains almost no plants

3 Apply the Skill

Directions: Choose the one best answer to each question.

Question 2 refers to the following information and graph.

Throughout the world, animals and plants face possible extinction. Some issues that threaten animals are habitat change, climate change, and disease. Some mammals at most serious risk are the tamaraw, a kind of water buffalo native to the Philippines; the Ethiopian wolf; the giant panda; and the Sumatran rhinoceros. The graph below shows population change of endangered mammals.

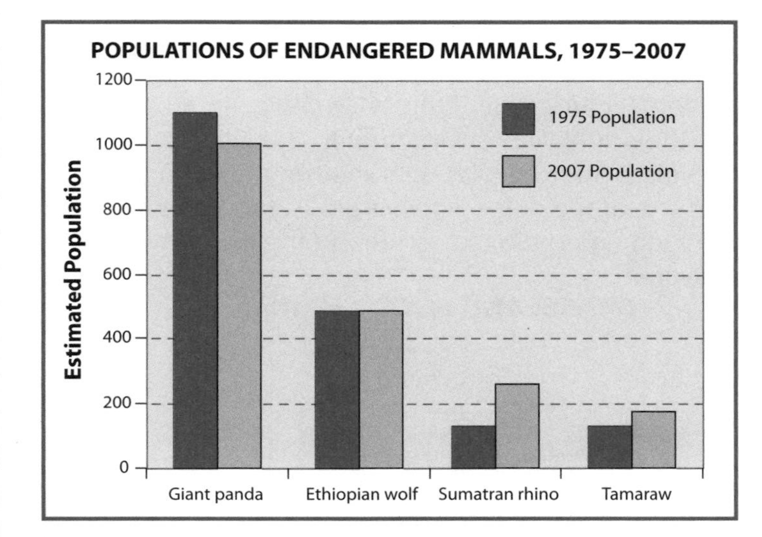

2. Which statement best describes what the bar graph shows?

   Between 1975 and 2007,

   (1) the populations of all endangered animals decreased
   (2) the populations of all endangered animals increased
   (3) the panda population decreased
   (4) the tamaraw population decreased
   (5) the wolf population increased

Questions 3 and 4 refer to the following information and graph.

In 1974, the gray wolf was listed as endangered under the Endangered Species Act. At that time, there were so few gray wolves left that scientists were not sure that the wolves would survive. The Endangered Species Act made it illegal for people to hunt gray wolves. The graph below shows the changes in the gray wolf population in Minnesota since the wolves were listed as endangered.

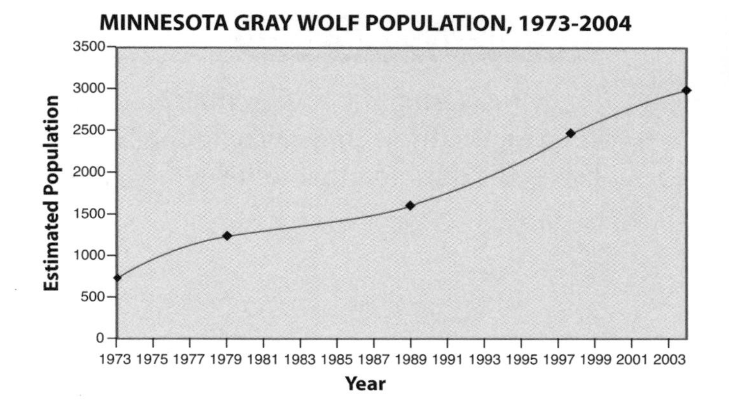

3. Based on the information in the paragraph and graph, which of the following statements is true?

   (1) Listing the gray wolf as endangered had no effect on its population size.
   (2) Gray wolves are still hunted today.
   (3) Gray wolves are nearly extinct.
   (4) The gray wolf population has increased under the protection of the Endangered Species Act.
   (5) Gray wolves have difficulty breeding in large numbers today.

4. Based on the information in the graph, in which year was the gray wolf population smallest?

   (1) 1973
   (2) 1979
   (3) 1989
   (4) 1998
   (5) 2004

UNIT 1

Interpret Diagrams

1 Learn the Skill

Diagrams show the relationships between ideas, objects, or events in a visual way. Diagrams also can be used to show the order in which events occur. When you **interpret diagrams**, you see how objects or events relate to one another.

2 Practice the Skill

By mastering the skill of interpreting diagrams, you will improve your study and test-taking skills, especially as they relate to the GED Science Test. Examine the diagram and strategy below. Then answer the question that follows.

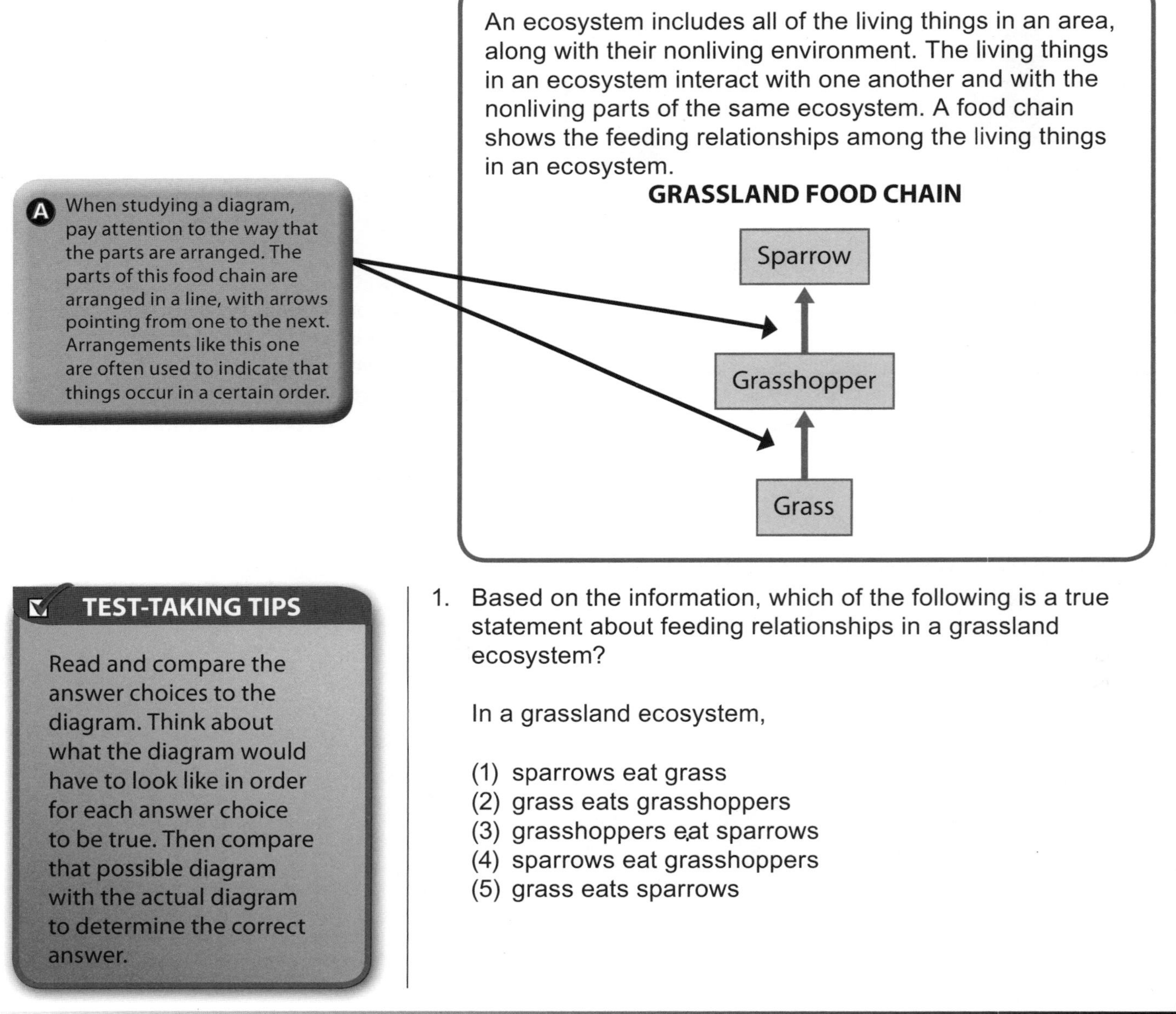

1. Based on the information, which of the following is a true statement about feeding relationships in a grassland ecosystem?

   In a grassland ecosystem,

   (1) sparrows eat grass
   (2) grass eats grasshoppers
   (3) grasshoppers eat sparrows
   (4) sparrows eat grasshoppers
   (5) grass eats sparrows

3 Apply the Skill

Directions: Choose the one best answer to each question.

Questions 2 through 4 refer to the following diagram.

DIETS OF TWO DESERT ANIMALS

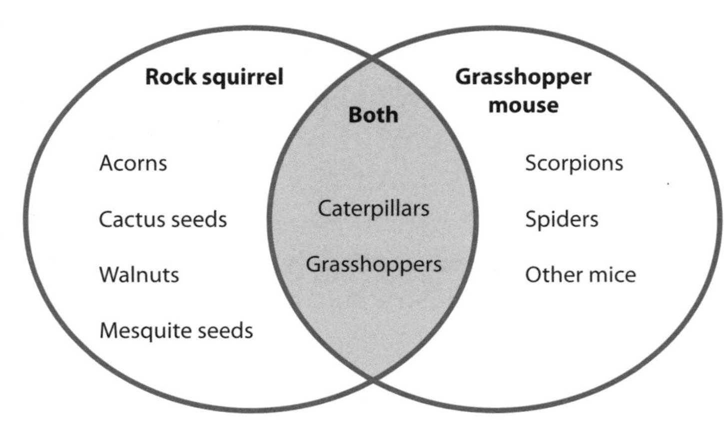

2. Based on the information in the diagram, which of these living things do both rock squirrels and grasshopper mice eat?

   (1) acorns
   (2) cactus seeds
   (3) scorpions
   (4) spiders
   (5) caterpillars

3. What can you learn from this diagram?

   (1) the foods eaten by some animals in a desert
   (2) which animals eat rock squirrels
   (3) the names of all the animals that live in the desert
   (4) which foods are most plentiful in the desert
   (5) where different organisms in an ecosystem live

4. Based on the diagram, which of the following is true?

   (1) Scorpions are prey for both rock squirrels and caterpillar mice.
   (2) Grasshopper mice eat other types of mice.
   (3) Only rock squirrels eat grasshoppers.
   (4) Acorns are higher in a food chain than walnuts.
   (5) Rock squirrels and grasshopper mice both eat spiders and caterpillars.

Questions 5 and 6 refer to the following information and diagram.

A food web is a more complete way to show the feeding relationships in an ecosystem. Unlike a food chain, which shows only one set of feeding relationships, a food web shows many different feeding relationships. Food webs show how many different living things are related.

GRASSLAND FOOD WEB

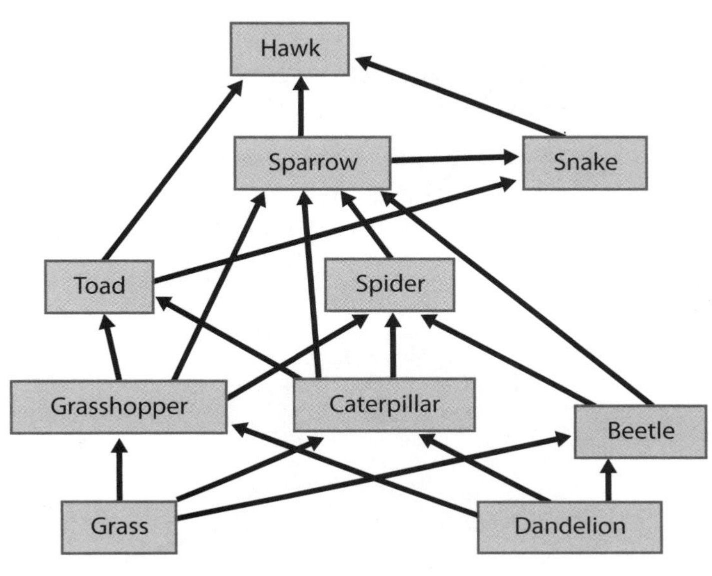

5. Based on the information in the diagram, what do sparrows eat?

   Sparrows eat

   (1) grass
   (2) plants and hawks
   (3) insects
   (4) plants and grasshoppers
   (5) hawks and snakes

6. Based on the information and diagram, which of the following statements is true?

   (1) Grass is food for many living things in a grassland.
   (2) Each animal eats only one other kind of animal or plant.
   (3) Hawks eat both plants and animals.
   (4) Food chains show more complex relationships than food webs.
   (5) Most animals in a grassland eat insects.

UNIT 1

LESSON 4

Interpret Illustrations

① Learn the Skill

Illustrations are often used to show many parts that make up a whole. For example, if an engineer is assembling an aircraft engine, he or she will look at several illustrations to see how the parts fit together. When you **interpret illustrations**, you look at the labels and figures within an illustration to understand how the parts of a whole fit together.

② Practice the Skill

By mastering the skill of interpreting illustrations, you will improve your study and test-taking skills, especially as they relate to the GED Science Test. Examine the illustration and read the strategies below. Then answer the question that follows.

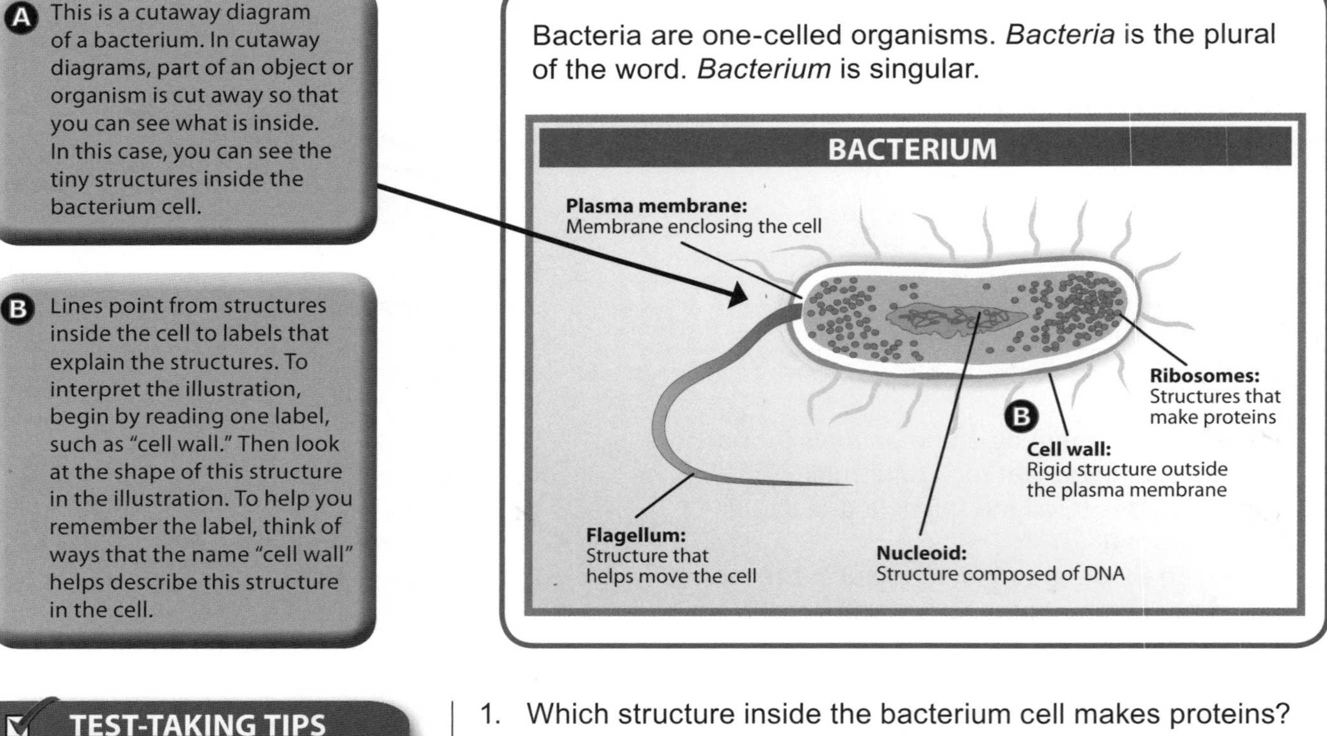

TEST-TAKING TIPS

Test questions may ask you to identify part of an illustration, or how parts of an illustration relate to one another. Look at the labels to determine these answers.

1. Which structure inside the bacterium cell makes proteins?

    (1) cell wall
    (2) flagellum
    (3) nucleoid
    (4) ribosomes
    (5) plasma membrane

③ Apply the Skill

Directions: Choose the one best answer to each question.

Questions 2 through 4 refer to the following information and illustration.

Animal cells all contain the same basic structures for making energy, digesting food, and reproducing. Mitochondria make energy, lysosomes digest food, and the beginning stages of reproduction occur in the nucleus. Some early light microscopes allowed scientists to see cells. But the electron microscope, invented in the 1950s, finally allowed scientists to see the internal structures of cells.

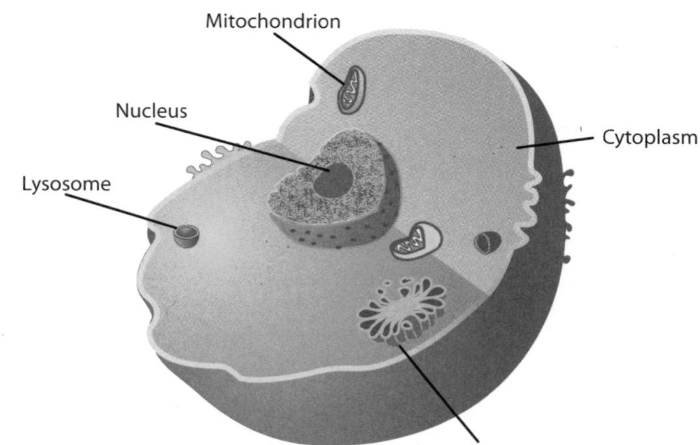

2. Based on the information and illustration, where are lysosomes located?

   (1) outside the cell
   (2) in the nucleus
   (3) in the Golgi complex
   (4) on the cell wall
   (5) in the cytoplasm

3. What would happen to this cell if something destroyed its nucleus?

   (1) It would not be able to reproduce properly.
   (2) It would have no energy for its processes.
   (3) It would store more energy.
   (4) It would become a plant cell.
   (5) It would become larger.

4. Which of following statements is true?

   Before the 1950s, scientists did not know

   (1) the shape of cells
   (2) the size of cells
   (3) that living things are made of cells
   (4) the location of ribosomes or lysosomes in cells
   (5) that cells use food and eliminate waste

Questions 5 and 6 refer to the following information and illustration.

Your body's many types of cells are organized into different systems, such as the circulatory system and the nervous system. Your nervous system is composed of about 100 billion nerve cells, or neurons. Neurons carry signals through your body that allow you to move, sense things, think, and learn. The illustration below shows a neuron.

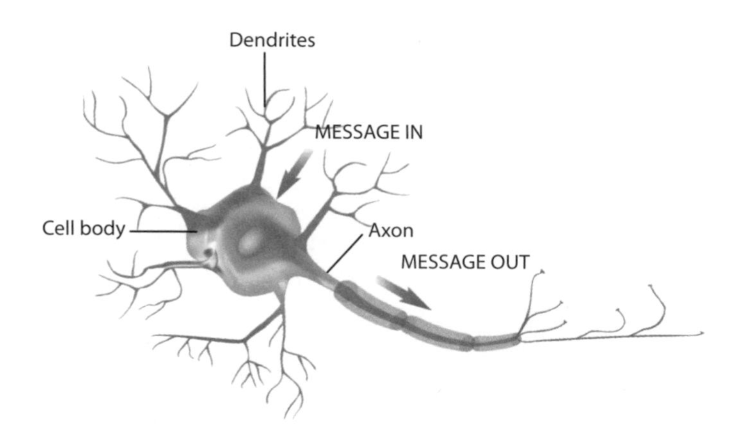

5. Based on the information and illustration, which of these statements is true of axons?

   (1) They are in the cell's nucleus.
   (2) They receive information.
   (3) They send information.
   (4) They are part of the cell body.
   (5) They hold the cell's hereditary material.

6. Based on the illustration, what function do dendrites serve?

   (1) They control growth.
   (2) They provide protection.
   (3) They produce energy.
   (4) They conduct electrical messages.
   (5) They aid in reproduction.

UNIT 1

LESSON 5

Main Idea and Details

1 Learn the Skill

The **main idea** is the most important point of a passage, story, table, or graphic. **Supporting details** provide additional information about the main idea. Such details include facts, statistics, explanations, and descriptions. A main idea may be clearly stated or it may be implied. If it is implied, you must use supporting details to determine the main idea.

2 Practice the Skill

By mastering the skill of determining the main idea and supporting details, you will improve your study and test-taking skills, especially as they relate to the GED Science Test. Read the paragraph and strategies below. Then answer the question that follows.

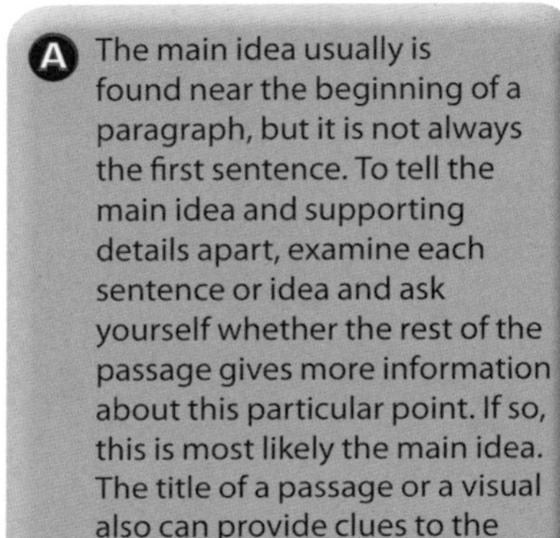

A The main idea usually is found near the beginning of a paragraph, but it is not always the first sentence. To tell the main idea and supporting details apart, examine each sentence or idea and ask yourself whether the rest of the passage gives more information about this particular point. If so, this is most likely the main idea. The title of a passage or a visual also can provide clues to the main idea.

B Supporting details generally follow the main idea. They give more information about the main idea. Here, the supporting details are examples of ways that many kinds of bacteria are helpful.

BACTERIA: NOT ALWAYS THE BAD GUY

We often think of bacteria in negative terms. Not all bacteria, however, cause disease. Soil may contain millions of bacteria. Many of these bacteria are decomposers, which break down and recycle nutrients in the bodies of dead plants and animals. Many helpful bacteria live inside the human body, where they aid digestion and produce vitamins the body needs. Bacteria play a role in the production of many popular foods such as cheeses. Scientists have even discovered how to put bacteria to work to clean up contaminated water and soil.

TEST-TAKING TIPS

Read all the answer choices for a question carefully. Just because an answer choice is a true statement does not mean it is the correct answer to the question.

1. Which of the following is another detail that supports the main idea of this passage?

   Bacteria

   (1) are single-celled organisms
   (2) are used in the production of some medications
   (3) are too small to see without a microscope
   (4) can be killed with antibiotics
   (5) can cause diseases such as strep throat

3 Apply the Skill

Directions: Choose the one best answer to each question.

Question 2 refers to the following table.

GROUPS OF BACTERIA BY SHAPE	
BACTERIA	**SHAPE**
Cocci	Sphere
Bacilli	Rod
Spirilla	Short, right corkscrew

2. Which main idea do the details of this table support?

   (1) Bacteria have many uses.
   (2) Rod-shaped bacteria are bacilli.
   (3) Bacteria can be grouped by structure.
   (4) All bacteria cause disease.
   (5) Cocci are the most common bacteria shape.

Question 3 refers to the following information.

The bacteria streptococci contain a variety of species, each with its particular effect. For example, *Streptococcus pyogenes*, or Group A streptococcus bacteria, can cause diseases from tonsilitis to scarlet fever. Other forms of streptococci can result in tooth decay, sinus infections, meningitis, or pneumonia. However, some forms of streptococci aid in the production of butter, yogurt, and certain cheeses.

3. Which detail supports the main idea that streptococci have both different species and effects?

   (1) *Streptococcus pyogenes* can cause diseases ranging from tonsilitis to scarlet fever.
   (2) Streptococci can cause sinus infections.
   (3) Some strains of streptococci cause life-threatening diseases such as pneumonia.
   (4) Tooth decay can be a result of a streptococcus bacterium.
   (5) Some forms of streptococci can cause disease, while others help produce foods.

Questions 4 and 5 refer to the following diagram and information.

Plants need nitrogen to survive and grow. However, most of the nitrogen in the environment is in a form (gaseous nitrogen) plants cannot use. Gaseous nitrogen can become a form of nitrogen that plants *can* use through a process called nitrogen fixation. In this process, certain types of bacteria in the soil "fix" nitrogen, or turn it into a different form. Some of these bacteria live in small nodules on the roots of some plants. Legumes, which include soybeans and peanuts, live with these nitrogen-fixing bacteria on their roots. Legumes are a major source of usable nitrogen in the soil.

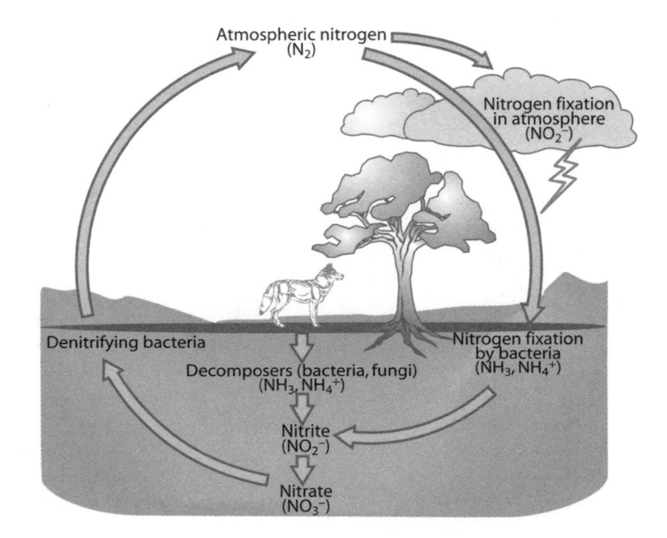

4. Which statement identifies the main idea of the passage?

   (1) Legumes include soybeans and peanuts.
   (2) Plants need nitrogen to survive.
   (3) Bacteria help provide nitrogen to plants.
   (4) Bacteria live in the root systems of legumes.
   (5) Plants cannot use gaseous nitrogen.

5. Together, the details in the passage and diagram support which main idea?

   (1) Nitrogen changes form as it is recycled in the environment.
   (2) Plants cannot survive without nitrogen-fixing bacteria.
   (3) Most nitrogen in the environment is in a form that plants cannot use.
   (4) Gaseous nitrogen is produced constantly in the atmosphere.
   (5) Plants use nitrogen they take in from the soil.

Summarize

1 Learn the Skill

When you **summarize** a passage or other feature such as a table, graph, or diagram, you briefly identify and describe its main points. A summary generally does not contain the exact words of the original passage, and it is almost always shorter.

2 Practice the Skill

By mastering the skill of summarizing, you will improve your study and test-taking skills, especially as they relate to the GED Science Test. Read the paragraph and strategies below. Then answer the question that follows.

A Summarizing involves separating the most relevant information (main idea and supporting details) from less relevant information (extra details). Here, the relative sizes of duckweed and sequoias is an interesting detail, but it does not belong in a summary. The data in the first sentence is also an extra, trivial detail.

B This paragraph does not state the main idea in a single sentence. Instead, there are two major points that form the main idea. A summary of this passage should identify these major points, but it should word them in a different, simpler way.

There are at least 300,000 different types of plants on Earth. Plants can be found in most types of environments—from cold polar areas to the heat of the tropics. Most plants live on land, but plants also grow in water. Tiny water plants, such as duckweed, have leaves just a few millimeters wide. At the other end of the size scale, giant sequoias can reach a height of more than 90 meters. Regardless of their size or the environment in which they live, plants play a critical role in the survival of many other organisms. Plants form the base of most food chains on Earth, capturing sunlight and using the energy to produce food in a process called photosynthesis. Oxygen, which plants give off during photosynthesis, is essential for animal life on Earth.

USING LOGIC

Summarizing requires using and recognizing synonyms or the different ways that an idea can be stated. Here, the passage does not state specifically that other organisms rely on plants, but all the details of the second half of the passage describe ways organisms depend, or rely, on plants.

1. Which sentence best summarizes the passage above?

   Plants

   (1) are a diverse group of organisms on which other organisms rely
   (2) produce oxygen that animals need
   (3) range in size from smaller than an inch to hundreds of feet tall
   (4) produce food for themselves and other forms of life during photosynthesis
   (5) live in almost every type of environment

3 Apply the Skill

Directions: Choose the one best answer to each question.

Questions 2 through 4 refer to the following information and diagram.

Plants can be grouped according to whether they have vascular tissue. Vascular tissue is a system of tubes running through a plant's body. One kind of vascular tissue, called xylem, carries water and minerals from the soil up through the stems and leaves. Xylem is made up of stacks of dead cells. The other kind of vascular tissue, called phloem, moves sugars produced in the leaves throughout the plant's body. Phloem is made up of living cells.

Nonvascular plants lack vascular tissues. Without such tissues to move materials throughout the plant's body, nonvascular plants must be small. This small size allows water and sugars to move through the plant cell by cell. Most nonvascular plants grow close to the ground in wet areas. Many of the cells of a nonvascular plant can take water and other nutrients directly from the environment.

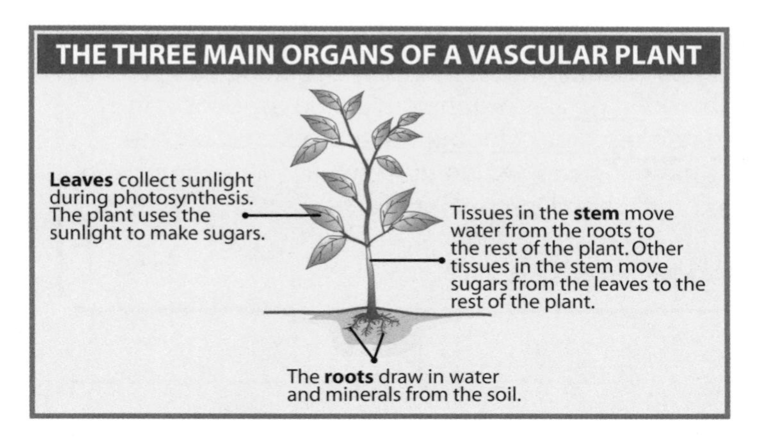

2. Which of the following is the most important idea in the diagram?

   (1) The organs of a vascular plant have specific roles.
   (2) Vascular plants have organs, whereas nonvascular plants do not.
   (3) Phloem is found in roots, stems, and leaves.
   (4) Vascular plants depend on roots to collect water.
   (5) Stems form the main part of a plant's body.

3. Which statement best summarizes the information in the passage?

   (1) Xylem carries water throughout a plant's body, and phloem carries sugars.
   (2) Nonvascular plants grow in wet areas.
   (3) Vascular plants are larger than nonvascular plants.
   (4) Plants can be grouped according to whether they have vascular tissue.
   (5) Vascular plants and nonvascular plants move materials through their bodies in different ways.

4. Which of the following points should be included in a summary of the passage and diagram?

   (1) Vascular plants have three basic organs: roots, stems, and leaves.
   (2) The organs of vascular plants work together to move materials through the plant's body.
   (3) Roots, stems, and leaves contain both xylem and phloem.
   (4) Both vascular plans and nonvascular plants carry out photosynthesis to produce sugars.
   (5) Most nonvascular plants grow close to the ground in wet areas.

Question 5 refers to the following information.

Cactuses are flowering plants that generally live in and have adapted to arid, or dry, regions. There are about 1,500 species of cactuses, the largest number of which grow in Mexico. Compared to other plants, cactuses have much fewer, if any, leaves. Without significant leaves, the thick, woody stems of the cactuses carry out photosynthesis. The root systems of cactuses are thick and shallow, allowing for broad absorption of surface moisture.

5. Which title best summarizes the passage?

   (1) Cactuses: Mexico's Underrated Flowers
   (2) The Unique Features of Cactuses
   (3) Life in Arid Regions
   (4) The Many Species of Cactuses
   (5) Root Systems and Water

LESSON 7

Use Context Clues

UNIT 1

1 Learn the Skill

Context clues can help you to determine the meaning of a word or passage. With context clues, you use details or restatements that surround the word or idea to determine its meaning.

2 Practice the Skill

By mastering the skill of context clues, you will improve your study and test-taking skills, especially as they relate to the GED Science Test. Read the paragraph and strategies below. Then answer the question that follows.

A When you use context clues to figure out the meaning of a word or phrase, examine familiar words to get an idea of the general tone of the passage. This passage contains many words and phrases, such as those underlined in red, that have negative meanings. Although *algal bloom* is not specifically defined in the passage, you can tell that it is a negative thing or event.

B Although the passage does not directly answer the question, you can use logic to figure it out. The passage states that nutrients can cause algal blooms and that farm fertilizers are a source of nutrients. From these clues, you can identify a solution to the problem of algal blooms.

"In addition to toxic pollutants, increased nutrients, especially nitrogen and phosphorus, from city sewage and fertilizers from agricultural areas (e.g., animal feed lots) have also proven to be very damaging to aquatic ecosystems. Certain levels of these nutrients are known to cause harmful algal blooms in both freshwater and marine habitats. In turn, algal blooms impact aquatic biodiversity by affecting water clarity, depleting oxygen levels, and crowding out organisms within an ecosystem. In some instances algal blooms have produced neuro-toxins that have led to species die-offs and illnesses such as paralytic shellfish poisoning."

From *Aquatic Biodiversity*, EPA, 2007

USING LOGIC

The word *bloom* often has a positive meaning, especially in the context of flowers blooming. You might assume that a bloom of any kind would be helpful in an ecosystem. A *bloom* in this context, however, is a sudden increase in population of algae, not a flowering.

1. Based on the information, which measure would help prevent algal blooms?

   An algal bloom could be prevented by

   (1) increasing the amount of oxygen in a body of water
   (2) decreasing the amount of fertilizer used on farms
   (3) developing a treatment for neurotoxins
   (4) adding nutrients to freshwater habitats
   (5) increasing the clarity of water

3 Apply the Skill

Directions: Choose the one best answer to each question.

Questions 2 and 3 refer to the following paragraph and diagram.

"Wetlands and riparian areas can play a critical role in reducing . . . pollution by intercepting surface runoff, subsurface flow, and certain ground water flows . . . Their role in water quality improvement includes processing, removing, transforming, and storing such pollutants as sediment, nitrogen, phosphorus, and certain heavy metals."

From *National Management Measures to Protect and Restore Wetlands and Riparian Areas*, EPA, 2005

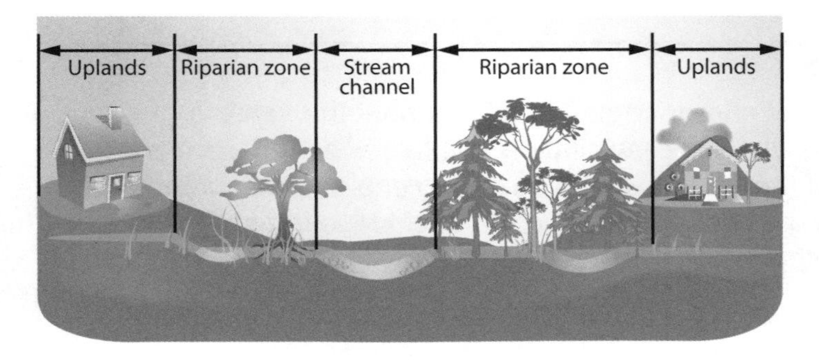

2. Based on the information and diagram, what is the best definition of a riparian area?

   A riparian area is

   (1) a heavily wooded area near farmland
   (2) a source of water pollution
   (3) the region directly surrounding a city
   (4) a zone of vegetation along a body of water
   (5) a freshwater stream

3. Based on the information, what argument is the EPA document most likely making?

   (1) Riparian areas should be preserved to decrease pollution of water resources.
   (2) Riparian areas should be eliminated because they are a major source of heavy metals.
   (3) Riparian areas should be replaced by wetlands to decrease water pollution.
   (4) Preserving riparian areas will involve destroying farmland.
   (5) Pollutants such as sediment, nitrogen, and phosphorus are destroying wetlands.

Question 4 refers to the following paragraph.

Common air pollutants can have harmful effects on the health and survival of plants and animals. The Clean Air Act (CAA) of 1970 established air quality monitoring for six criteria pollutants in the United States. Leaders in states where the criteria pollutants are above levels established in the CAA must submit plans to bring those levels down. The federal government can issue sanctions against a state that fails to comply with the law. In general, the CAA has helped reduce the level of these air pollutants over the last 30 years.

4. Based on the information, what is a "criteria pollutant"?

   A criteria pollutant is a pollutant that

   (1) is no longer a threat to the health of plants and animals
   (2) is common and must be tested regularly
   (3) is generally rare
   (4) does not cause health problems in humans
   (5) is easiest to monitor

LESSON 8

Analyze Information

UNIT 1

1 Learn the Skill

Graphical information, such as graphs, tables, and illustrations, often is used to support and clarify text. Graphics also may be used to present additional information. When you see text and graphical information together, you must **analyze information** in the text and the graphic to fully understand the material being presented.

2 Practice the Skill

By mastering the skill of analyzing information, you will improve your study and test-taking skills, especially as they relate to the GED Science Test. Examine the text and graph. Then answer the question that follows.

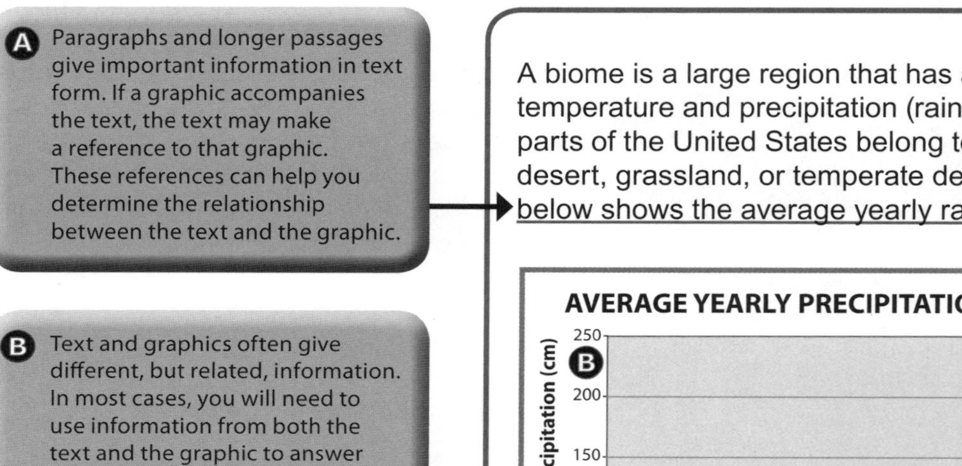

A Paragraphs and longer passages give important information in text form. If a graphic accompanies the text, the text may make a reference to that graphic. These references can help you determine the relationship between the text and the graphic.

B Text and graphics often give different, but related, information. In most cases, you will need to use information from both the text and the graphic to answer the questions.

A biome is a large region that has about the same temperature and precipitation (rainfall) everywhere. Most parts of the United States belong to one of three biomes: desert, grassland, or temperate deciduous forest. The graph below shows the average yearly rainfall for several biomes.

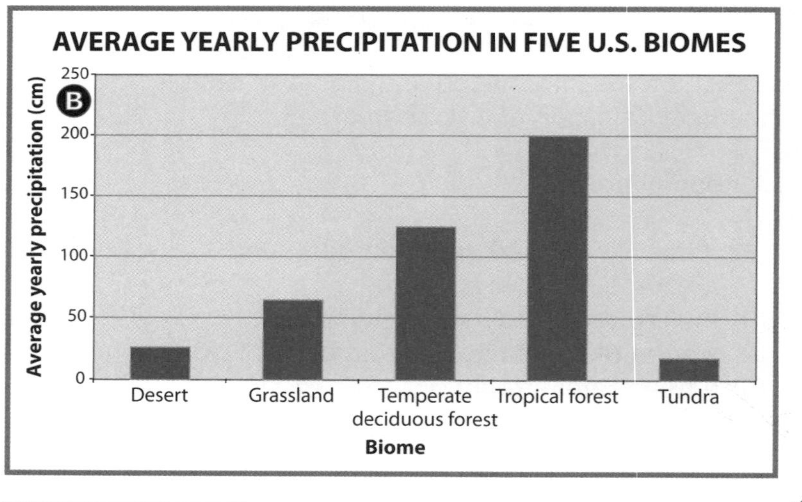

USING LOGIC

In many cases, the correct answer is not specifically stated in the text or shown in the graphic. Instead, you must combine information from both to answer the question.

1. Based on the information and graph, which of the following statements is true?

   (1) Most of the United States is part of the grassland biome.
   (2) Most parts of the United States receive between 25 cm and 125 cm of rainfall each year.
   (3) The desert is the driest biome on Earth.
   (4) Tropical forests and grasslands receive about the same amount of rainfall each year.
   (5) The three wettest biomes are found in the United States.

3 Apply the Skill

Directions: Choose the one best answer to each question.

Questions 2 through 4 refer to the following paragraph and graph.

The plants and animals that live in a particular biome have adaptations to help them survive that biome's climate. These adaptations may take many forms. Some adaptations are physical characteristics. For example, plants that live in dense forests may grow very tall, allowing them to get sunlight by growing above other plants in the forest. Other adaptations involve behaviors. For example, animals that live in a biome with cold winters may hibernate or migrate to a warmer area during the winter.

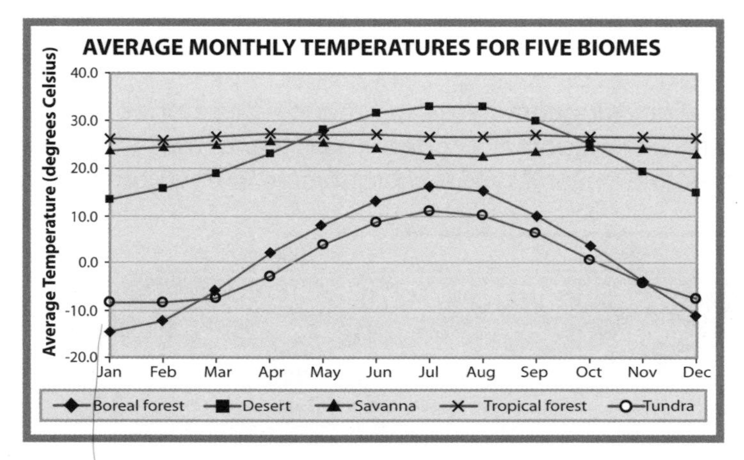

2. Based on the information and graph, animals in which biomes are most likely to have thick fur to help them stay warm?

   (1) savanna and desert
   (2) boreal forest and tropical forest
   (3) desert and tundra
   (4) tundra and boreal forest
   (5) tropical forest and savanna

3. Based on the information and graph, which of the following is true?

   (1) Plants in the desert grow very tall.
   (2) The average temperature in a savanna is warmer than in a tropical forest.
   (3) Plants and animals will adapt differently depending on the biome in which they live.
   (4) Temperatures in the tundra are always colder than in a boreal forest.
   (5) Animals in a tropical forest are likely to have fur.

4. Based on the information and graph, animals living in the boreal forest and tundra biomes are most likely to have which of the following characteristics?

   (1) large ears to rid excess body heat
   (2) fur that changes color to provide protection from predators
   (3) the ability to hibernate during very cold weather
   (4) a great deal of exposed skin to increase the body's ability to stay cool
   (5) a reliance on heat from the environment to stay warm

Question 5 refers to the following paragraph and map.

The main factors that define a biome are temperature and precipitation. Regions that have similar temperatures and amounts of precipitation are generally placed into the same biome, even if they are very far apart geographically. The map shows the distribution of biomes in the world.

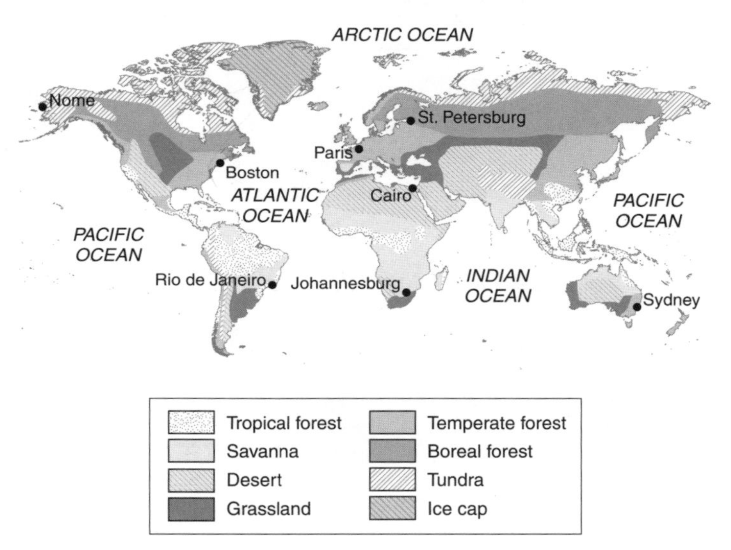

5. Based on the information in the paragraph and map, which two cities are most likely to have similar temperatures and amounts of precipitation?

   (1) Boston and Paris
   (2) Nome and Cairo
   (3) Sydney and St. Petersburg
   (4) Boston and Rio de Janeiro
   (5) St. Petersburg and Johannesburg

Categorize and Classify

1 Learn the Skill

When you **categorize**, you choose the criteria for placing organisms, objects, or processes into groups. Such groups are based on common features or relationships between things in the group. When you **classify**, you put things into groups that already exist.

2 Practice the Skill

By mastering the skills of categorizing and classifying, you will improve your study and test-taking skills, especially as they relate to the GED Science Test. Read the table and strategies below. Then answer the question that follows.

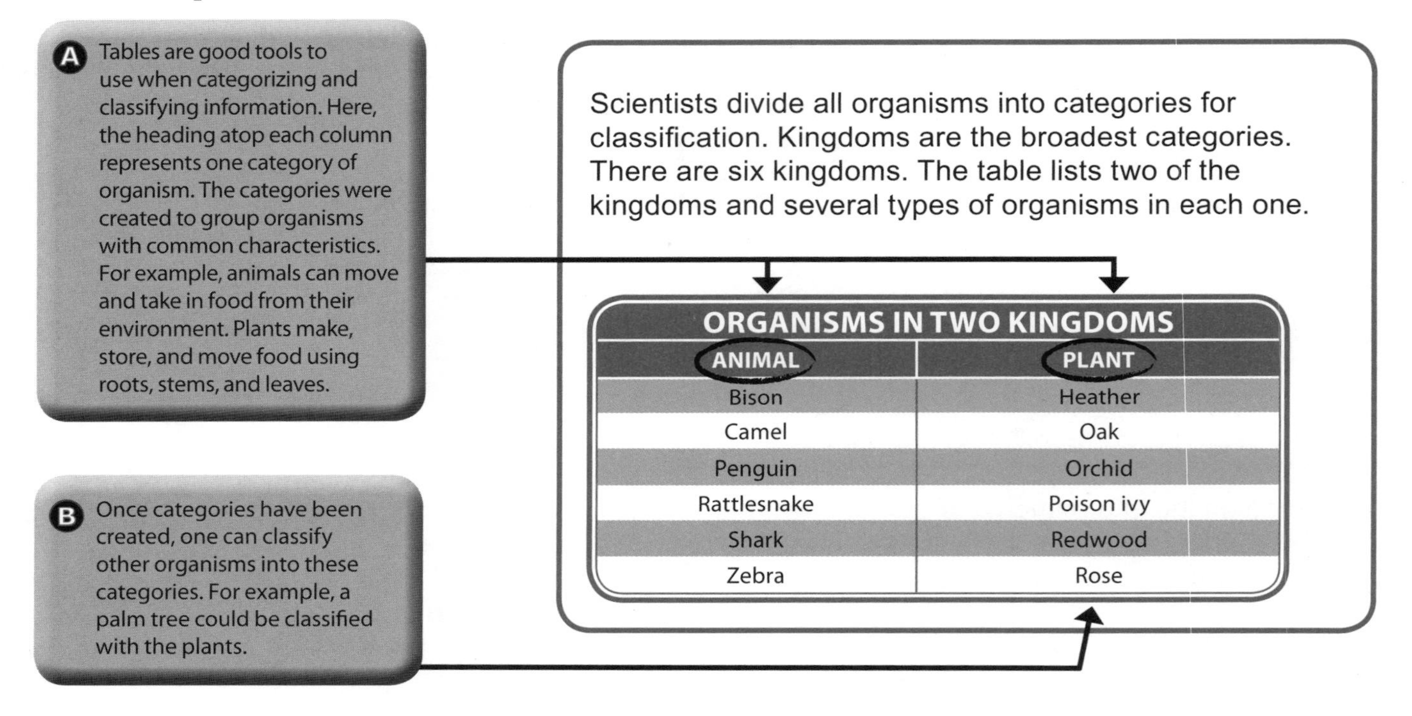

Scientists divide all organisms into categories for classification. Kingdoms are the broadest categories. There are six kingdoms. The table lists two of the kingdoms and several types of organisms in each one.

ORGANISMS IN TWO KINGDOMS

ANIMAL	PLANT
Bison	Heather
Camel	Oak
Penguin	Orchid
Rattlesnake	Poison ivy
Shark	Redwood
Zebra	Rose

TEST-TAKING TIPS

There is usually more than one way to categorize and classify organisms. If a question asks you to categorize, make sure to choose criteria that fit all the members of that group. If it asks you to classify, make sure the organism meets all criteria for that category.

1. Which phrase best describes the information in the table?

   (1) different types of bison and heather
   (2) an explanation of how plants and animals differ
   (3) animals and plants that are similar
   (4) two groups of organisms with different characteristics
   (5) the six kingdoms that classify living things

③ Apply the Skill

Directions: Choose the one best answer to each question.

Question 2 refers to the following passage.

In 1735, botanist Carolus Linnaeus created a system for categorizing and classifying organisms. Linnaeus categorized organisms by their physical traits. The broadest category is the kingdom. Kingdoms are divided into smaller groups called phyla (singular, *phylum*). Phyla are divided into even smaller categories called classes, and so on down to the smallest category, species. A species is one particular type of animal—a category of one. One kingdom can have millions of different species.

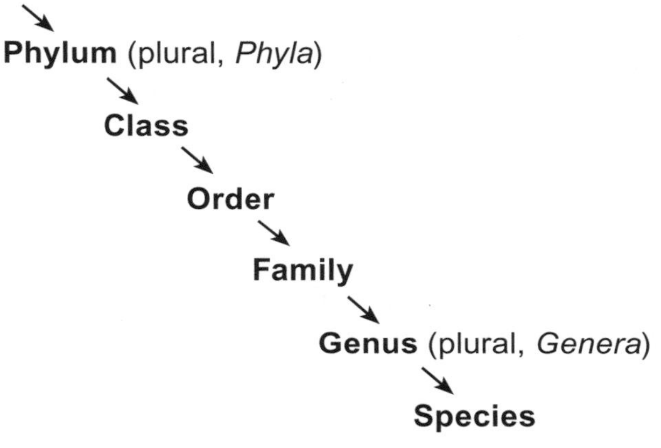

2. Based on the information, which statement best describes the structure of the Linnaean classification system?

   The Linnaean classification system has

   (1) more families than classes
   (2) fewer species than genera
   (3) the same number of phyla and classes
   (4) more kingdoms than species
   (5) more kingdoms than any other category

Question 3 refers to the following paragraph and table.

The six kingdoms that scientists use to classify all living things are in the far left column of the table. Categories of information about each kingdom are in the other columns.

SIX SCIENTIFIC KINGDOMS

Kingdom	Characteristics of Organisms	Examples
EUBACTERIA	Single-celled; some species form chains or mats	*E. coli*, which is found in the human digestive system
ARCHAEBACTERIA	Single-celled; many can survive in extreme environments	Organisms that live in hot springs
PROTISTA	Some single-celled, some multicellular; some make their own food, some eat other organisms	Paramecia, amoebas, algae
FUNGI	Most are multicellular; made of threadlike fibers	Yeasts, molds, mushrooms
PLANTAE	Contain many specialized cells; cells have rigid outer wall; feature structures to make own food	Mosses, ferns, flowering plants
ANIMALIA	Made up of many specialized cells; cells lack rigid outer wall; must collect and eat food	Sponges, worms, insects, fish, amphibians, reptiles, birds, mammals

3. A scientist discovers a new organism and classifies it in kingdom Plantae. Based on the information in the table, what must be true about the organism?

   The organism

   (1) has threadlike fibers
   (2) can survive in extreme environments
   (3) can produce its own food
   (4) is made up of a single cell
   (5) is more similar to bacteria than to animals

Compare and Contrast

1 Learn the Skill

When you **compare**, you identify the ways in which organisms, objects, places, or events are similar and the ways in which they are different. When you **contrast**, you identify only how those organisms, objects, places, or events are different.

2 Practice the Skill

By mastering the skills of compare and contrast, you will improve your study and test-taking skills, especially as they relate to the GED Science Test. Examine the illustrations and strategies below. Then answer the question that follows.

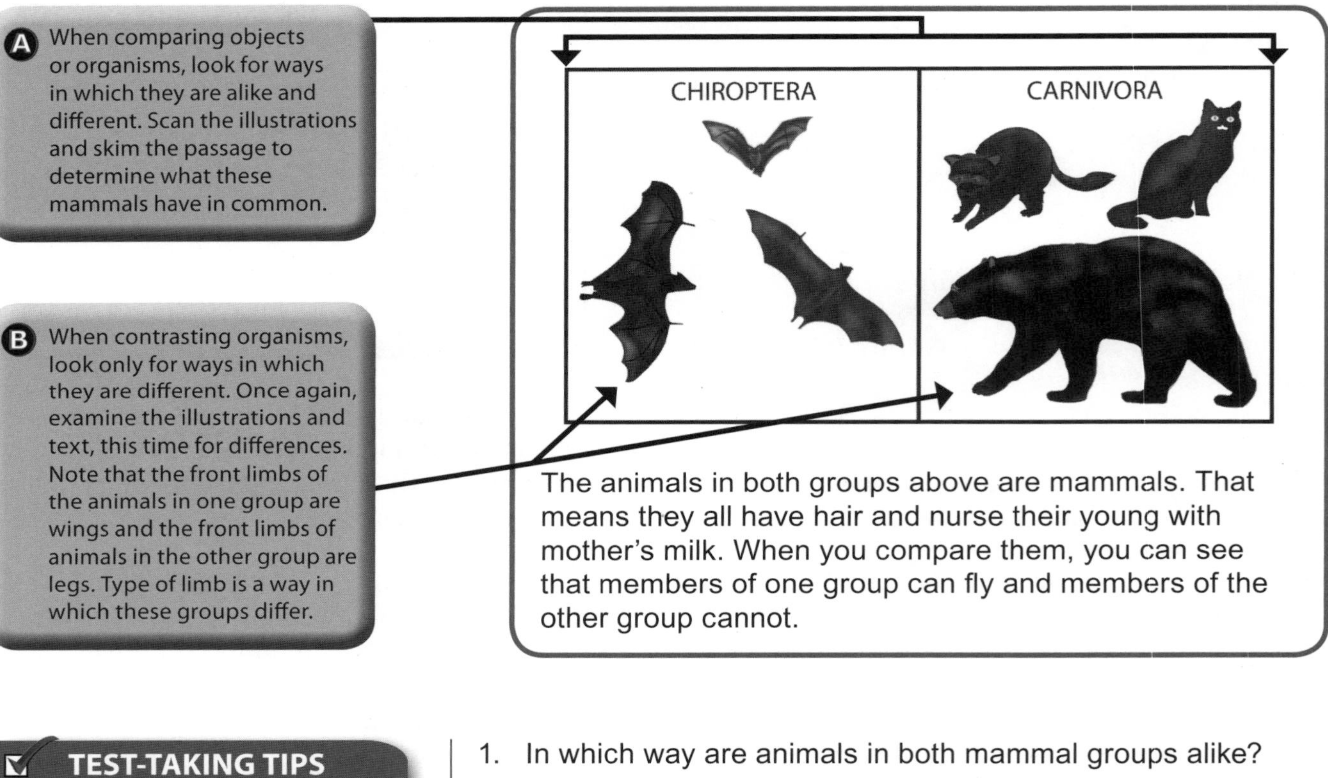

TEST-TAKING TIPS

Words such as *both, as, similar, like,* and *alike* often show comparisons. Words such as *unlike, different,* and *but* often show contrasts. Consider possible similarities and differences before reading the answer choices. Then look for the correct answer.

1. In which way are animals in both mammal groups alike?

   They

   (1) lay eggs in which young develop
   (2) have the same type of limb
   (3) have scales for skin
   (4) are the same size
   (5) have hair

③ Apply the Skill

Directions: Choose the one best answer to each question.

Questions 2 and 3 refer to the following table.

THREE CLASSES OF MAMMALS		
MONOTREMES	**MARSUPIALS**	**PLACENTAL MAMMALS**
Young hatch from eggs	Young born live	Young born live
Mothers make milk	Mothers make milk	Mothers make milk
After hatching, young nurse by licking milk the mother squirts on her fur	Young suck mother's milk while developing inside her pouch	Young suck mother's milk after developing inside her uterus
Warm-blooded	Warm-blooded	Warm-blooded
Hair or fur	Hair or fur	Hair or fur
Four-chambered hearts	Four-chambered hearts	Four-chambered hearts
Relatively large brains	Relatively large brains	Relatively large brains
Live only in Australia and New Guinea	Live mainly in Australia and New Zealand	Live in almost all parts of the world

2. Compare these three classes of mammals. Which characteristic do they all share?

   (1) They live in Australia and New Zealand.
   (2) Their young hatch from eggs.
   (3) Mothers produce milk for their young.
   (4) Young mammals are born live.
   (5) Young mammals develop in a pouch.

3. Based on the table, which statement best contrasts the three classes of mammals?

   Unlike marsupials and placental mammals, the monotremes

   (1) have hair or fur
   (2) produce milk for their young
   (3) have young that develop in a uterus
   (4) are warm-blooded
   (5) lay eggs

Questions 4 and 5 refer to the following illustrations and table.

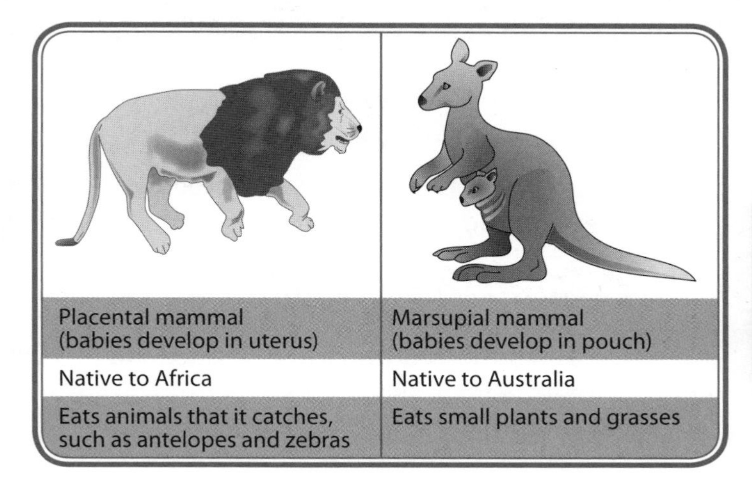

Placental mammal (babies develop in uterus)	Marsupial mammal (babies develop in pouch)
Native to Africa	Native to Australia
Eats animals that it catches, such as antelopes and zebras	Eats small plants and grasses

4. Which statement accurately contrasts these two animals?

   (1) Both the kangaroo and the lion are mammals.
   (2) The lion is a hunter, whereas the kangaroo is not.
   (3) Neither the kangaroo nor the lion is native to North America.
   (4) The lion and the kangaroo both have hair on their bodies.
   (5) The lion has live babies, whereas the kangaroo does not.

5. Based on the illustration and table, which statement accurately compares kangaroos with lions?

   (1) Both are monotremes.
   (2) Both graze on large prairies.
   (3) Both give birth to live young.
   (4) Both walk exclusively on four legs.
   (5) Both are native to Africa.

LESSON 11

Sequence

UNIT 1

1 Learn the Skill

When you place events or processes in **sequence**, you put them in a particular order. The sequence of events shows their relationship to, and with, one another.

2 Practice the Skill

By mastering the skill of sequencing, you will improve your study and test-taking skills, especially as they relate to the GED Science Test. Examine the diagram and strategies below. Then answer the question that follows.

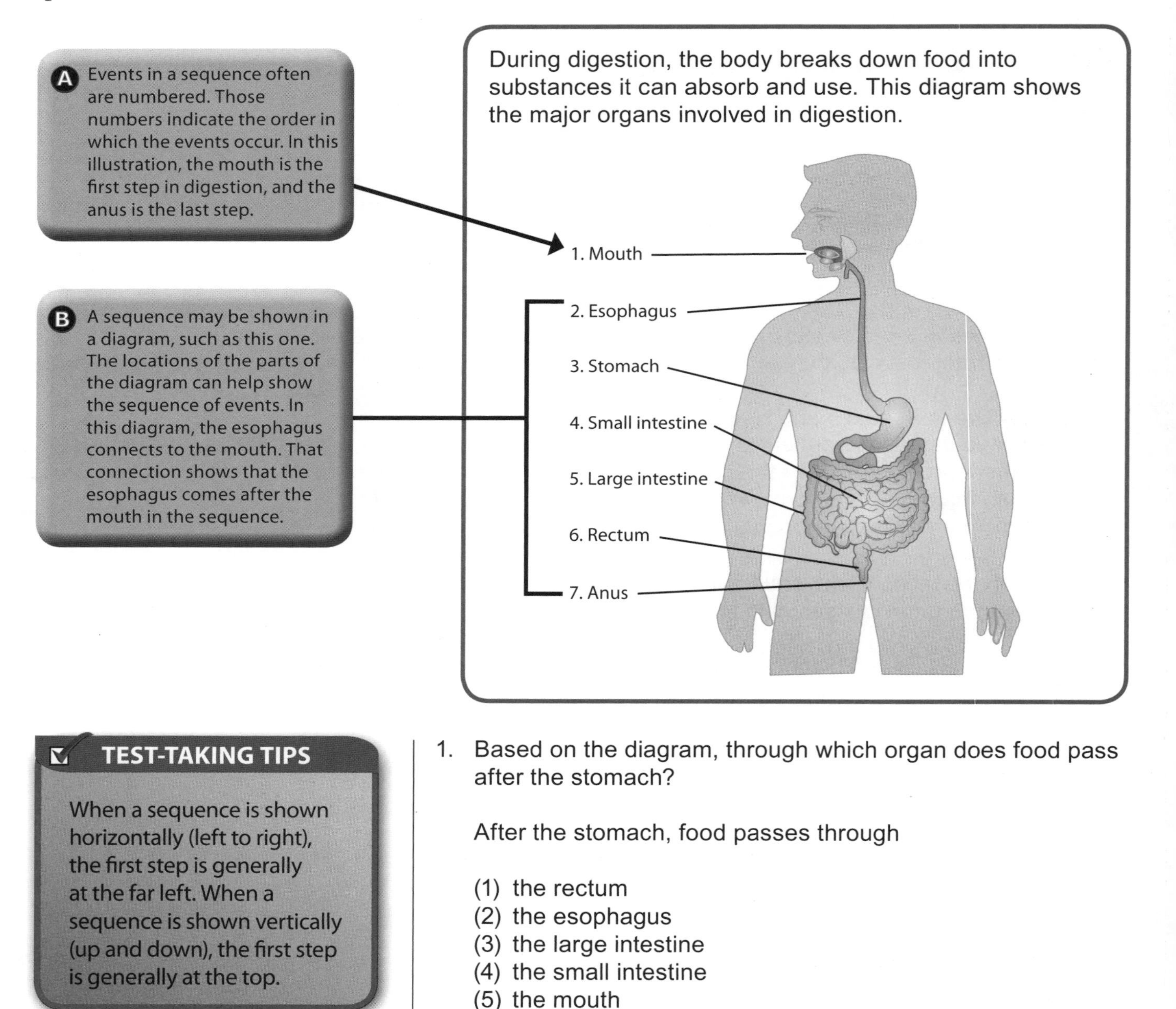

TEST-TAKING TIPS

When a sequence is shown horizontally (left to right), the first step is generally at the far left. When a sequence is shown vertically (up and down), the first step is generally at the top.

1. Based on the diagram, through which organ does food pass after the stomach?

   After the stomach, food passes through

   (1) the rectum
   (2) the esophagus
   (3) the large intestine
   (4) the small intestine
   (5) the mouth

③ Apply the Skill

Directions: Choose the one best answer to each question.

Questions 2 and 3 refer to the following paragraph.

Digestion actually begins before you take the first bite of food. The sight of a piece of warm pie and its tempting smell cause your salivary glands to send saliva into your mouth. Substances in saliva start digestion when food enters your mouth. Once you take a bite, your teeth cut, grind, and mash the food into smaller pieces that are easier to swallow. The substances in your saliva that aid digestion then can break down the food more easily. Your tongue also plays a part. You taste food with your tongue. But your tongue also moves the food around, forms it into a ball, and then pushes that ball to the back of your mouth.

2. When does the body produce saliva?

   The body first produces saliva when you

   (1) put food into your mouth
   (2) take a bite of food
   (3) see and smell the food
   (4) swallow the food
   (5) taste the food

3. Which list shows the correct sequence of steps in part of the digestive process?

   (1) teeth grind up food; saliva breaks down food; teeth move food to back of mouth
   (2) food moves to back of mouth; food moves to stomach; saliva released into mouth
   (3) teeth grind food; food enters mouth; saliva enters mouth
   (4) food enters mouth; tongue moves food to back of mouth; teeth grind food
   (5) teeth grind food; tongue moves food to back of mouth; saliva enters mouth

Questions 4 and 5 refer to the following text and diagram.

After food leaves the mouth, it passes into the esophagus. From there, it passes through the following sequence of processes.

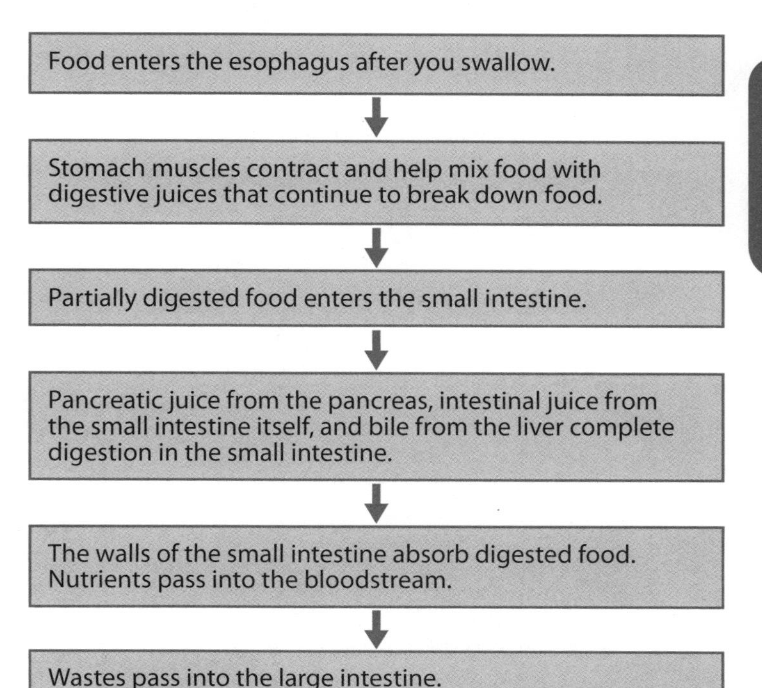

4. Through which organ do substances from the pancreas and liver enter the digestive tract?

   (1) small intestine
   (2) stomach
   (3) large intestine
   (4) esophagus
   (5) mouth

5. Based on the information, at what point is digestion mostly complete?

   Digestion is mostly complete when the food

   (1) leaves the esophagus
   (2) reaches the liver
   (3) leaves the small intestine
   (4) leaves the stomach
   (5) enters the stomach

LESSON 12

Interpret Timelines

1 Learn the Skill

Timelines display a sequence of events in a visual way. Remember that when you use a sequence, you place events or processes in a particular order. When you **interpret timelines**, you use a visual tool to determine when events happened and the amount of time that occurred between them. Timelines may be organized either horizontally or vertically.

2 Practice the Skill

By mastering the skill of interpreting timelines, you will improve your study and test-taking skills, especially as they relate to the GED Science Test. Read the text, timeline, and strategies below. Then answer the question that follows.

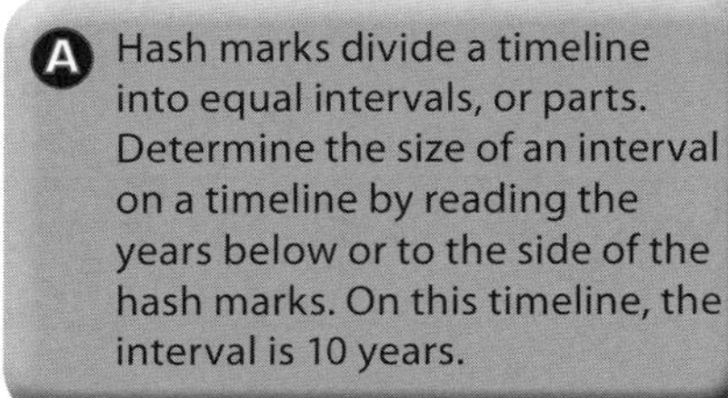

B When events occur at hash marks, the year is easy to identify. Here, you can see the Lacey Act was passed in 1900. When events occur between hash marks, you must estimate the year. The ESA was passed *between* 1970 and 1980. Based on where the event is marked, you can estimate that the year was about 1973.

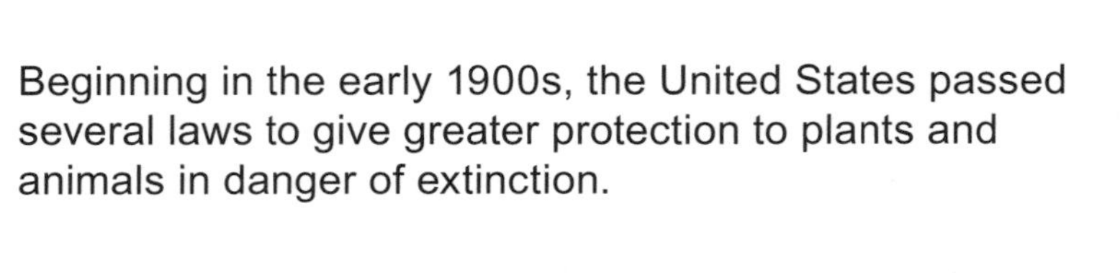

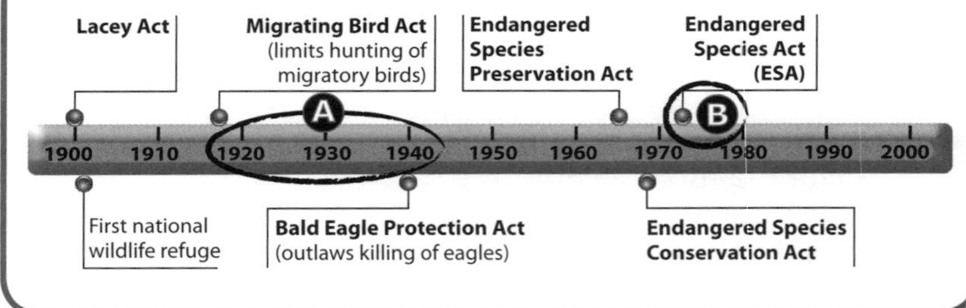

MAKING ASSUMPTIONS

A timeline can help you see trends in a sequence of events. For example, a timeline may show that the government passed few laws to protect plants and animals in the 1980s. This trend may support several ideas. Perhaps existing laws were working and no new laws were needed.

1. Which idea does the timeline support?

   (1) Protection of wildlife was more important to people in 1900 than in 1973.
   (2) National Wildlife Refuges were created as part of the Endangered Species Act.
   (3) Protection of endangered species was still a concern in the second half of the 1900s.
   (4) The Migrating Bird Act replaced the Bald Eagle Protection Act.
   (5) The Lacey Act provided effective protection to endangered species.

③ Apply the Skill

Directions: Choose the one best answer to each question.

Question 2 refers to the following information and timeline.

Sometimes legal protection alone cannot save a species from extinction. The timeline shows several species that became extinct during the 1900s.

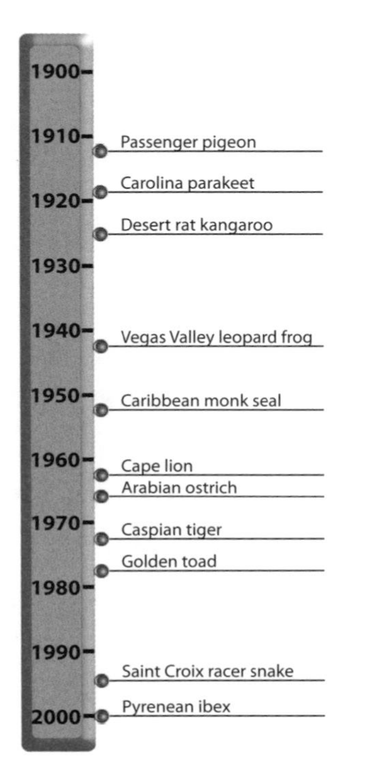

2. Based on the timeline, which statement is accurate?

   (1) The Caribbean monk seal became extinct in the 1960s.
   (2) Most of the animals listed became extinct in the first half of the 1900s.
   (3) Ten years passed between the extinctions of the Caspian tiger and the golden toad.
   (4) The timeline shows several extinctions between 2000 and present day.
   (5) The Cape lion, Arabian ostrich, and Caspian tiger became extinct within ten years.

Questions 3 and 4 refer to the following paragraph and timeline.

When the bald eagle became a national symbol in 1782, there were about 500,000 bald eagles in the United States. Hunting and habitat destruction, however, caused gradual declines in eagle populations. In the first part of the 1900s, the United States government gave eagles limited protection. Between the 1940s and 1960s, increased use of the pesticide DDT caused reproductive problems for eagles. As a result, their population numbers dropped. By 1963, the population of eagles was at an all-time low. The government passed laws in the 1960s and 1970s to give eagles greater protection.

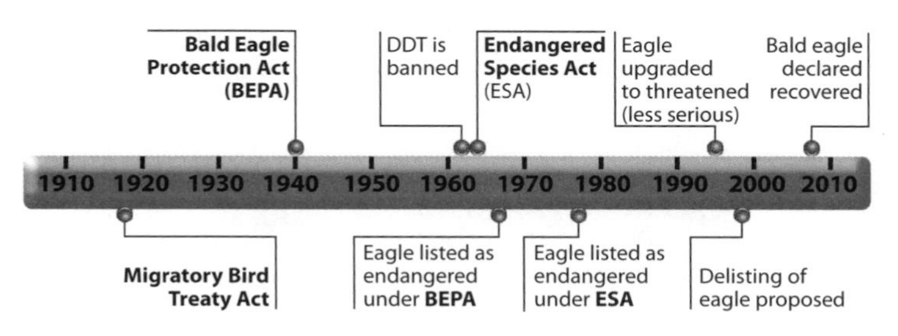

3. Based on the information and timeline, which statement was true of bald eagles in the 1960s?

   (1) They were abundant in the wild.
   (2) Their populations had remained virtually unchanged for nearly 200 years.
   (3) They were removed from the endangered list.
   (4) They produced fewer chicks because of DDT.
   (5) Their recovery was due to the Bald Eagle Protection Act.

4. Based on the information, which statement best describes the effect that federal actions had on bald eagle populations after 1950?

   (1) Negative effect; protections hastened the decline of bald eagles.
   (2) No effect; bald eagles never required government protection.
   (3) Minor effect; the population of bald eagles seldom changed over a 90-year period.
   (4) Moderate effect; limited government protection kept the populations stable.
   (5) Major effect; protections allowed for the removal of eagles from the endangered list.

Cause and Effect

1 Learn the Skill

A **cause** is an object or action that makes an event happen. The **effect** is the event that results from the cause. Often, a cause is stated directly. In other cases, it is implied. A cause may have more than one effect, and a single effect may have more than one cause.

2 Practice the Skill

By mastering the skill of cause and effect, you will improve your study and test-taking skills, especially as they relate to the GED Science Test. Look at the diagram and strategies below. Then answer the question that follows.

A Here, one cause can trigger several effects. The cause is an object that cuts the skin and inserts bacteria. The initial effect is that the bacteria cause the release of chemicals such as histamines.

B The histamines then become the cause of a secondary effect—the swelling and redness around the wound.

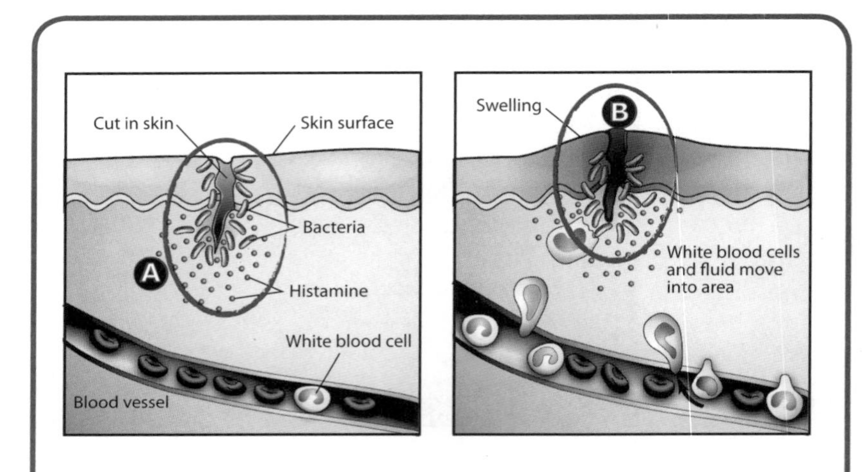

When a sharp object breaks the skin, bacteria can enter the body. Bacteria trigger a part of the body's immune system. As a first step to fighting infection, the body releases a chemical called histamine.

USING LOGIC

Just because events occur at or around the same time does not ensure a cause-and-effect relationship. When answering questions on cause and effect, consider whether it is logical that a given object or event influenced the other.

1. What is the body's response to this chemical signal?

   The body

   (1) drains blood and fluids from the injured area
   (2) sends in its own bacteria to attack the foreign bacteria
   (3) increases the flow of blood and other fluids to the area, which produces swelling
   (4) decreases the amount of oxygen in the area
   (5) surrounds the wound with iron from the blood

3 Apply the Skill

Directions: Choose the one best answer to each question.

Question 2 refers to the paragraph and graphic organizer.

In the late 1700s, the deadly disease smallpox was widespread. Edward Jenner, a doctor, found that most people who caught a related but milder disease called cowpox recovered. He further discovered that later exposure to smallpox did not affect these people. Jenner decided to expose a healthy young boy to the cowpox virus. The boy became mildly ill and then recovered. Jenner then exposed the boy to smallpox. The boy did not become ill. Without knowing it, Jenner had given the first vaccination. Vaccination involves injecting people with dead or weakened bacteria or viruses. As a result, a person's immune system produces antibodies to a particular bacterium or virus. Antibodies are proteins that identify disease-causing invaders. If a person is later exposed to the same microbe, the antibodies from his or her immune system can quickly recognize the invader. Today, vaccination is a major weapon in preventing disease.

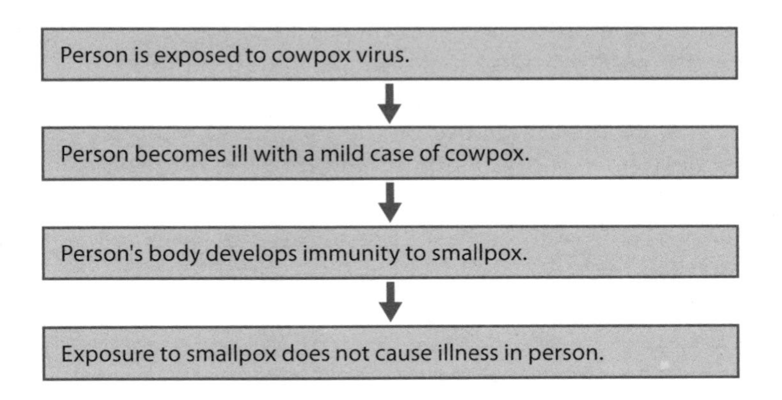

2. Based on the graphic organizer, what does the body produce as a result of exposure to cowpox?

    (1) more cowpox virus
    (2) antibodies that protect against smallpox
    (3) red blood cells that protect against antibodies
    (4) bacteria that cause cowpox
    (5) antibodies that protect against any disease

Question 3 refers to the paragraph.

The body's immune system responds when microbes such as viruses and bacteria invade. When viruses attack cells, the cells in turn produce proteins called interferons. These interferons spread to other cells, helping them resist the spread of the virus. The presence of foreign bacteria also causes the activation of certain proteins in the blood, which then destroy the invading microbes. Some of these proteins also trigger the inflammatory response that causes swelling and redness in the area of a wound.

3. Which of the following is a response of the immune system to invasion by bacteria?

    (1) It produces viruses.
    (2) It produces poisons that kill infected cells.
    (3) It increases the body's supply of blood.
    (4) It fills all the body's cells with protein.
    (5) It sends proteins in the blood to destroy bacteria.

Question 4 refers to the paragraph.

It wasn't until the 1850s and the efforts of scientist Louis Pasteur that people began to understand the cause of disease. Pasteur correctly believed that disease was caused by microscopic germs that attacked the body from outside. However, many scientists of that time thought it ridiculous that unseen organisms could destroy vastly larger ones. Pasteur's discovery, known as the germ theory of disease, convinced them otherwise. Pasteur later used his germ theory as a basis to explain the causes of and treatments for diseases such as anthrax and rabies. Later scientists built upon Pasteur's theory with the development and treatment of various types of vaccines.

4. What effect did Pasteur's research have on the understanding and prevention of disease?

    (1) Negative effect; his findings led to increased levels of disease in humans.
    (2) No effect; levels of disease remained unchanged.
    (3) Minor effect; levels of disease decreased, but only slightly.
    (4) Moderate effect; Pasteur's findings worked on certain diseases, but not others.
    (5) Major effect; Pasteur's efforts changed how people understand and treat diseases.

Make Inferences

UNIT 1

1 Learn the Skill

An **inference** is a logical guess based on facts, evidence, experience, or reasoning. If you know several related facts, you can **make inferences** about their meaning. That meaning could be implied, or hinted at, through several details. When meaning is implied, the reader must use prior knowledge to connect facts.

2 Practice the Skill

By mastering the skill of making inferences, you will improve your study and test-taking skills, especially as they relate to the GED Science Test. Read the excerpt and the strategies below. Then answer the question that follows.

A Before the author made an inference, he collected and presented his observations. Making an inference requires the use of logic to determine the truth based on available information.

B Scientists generally make inferences about objects or events that they cannot or did not observe directly. Darwin did not know from direct observation that small eyes help prevent infection. Instead, Darwin made an inference from the evidence he had.

Excerpt fom Charles Darwin's *On the Origin of Species by Means of Natural Selection* (1859)

"....The eyes of moles and of some burrowing rodents are rudimentary in size, and in some cases are quite covered up by skin and fur. This state of the eyes is probably due to gradual reduction from disuse . . . In South America, a burrowing rodent, the tuco-tuco . . . is even more subterranean in its habits than the mole; and . . . they were frequently blind; one which I kept alive was certainly in this condition, the cause, as appeared on dissection, having been inflammation of the nictitating membrane. As frequent inflammation of the eyes must be injurious to any animal, and <u>as eyes are certainly not indispensable to animals with subterranean habits, a reduction in their size . . . might in such case be an advantage</u>. . . ."

USING LOGIC

As you analyze a passage, note that a writer generally will state his or her inference after presenting all the facts. Inferences often can be recognized by the following wording: *if . . . then; as . . . then; might be; could be; probably.*

1. Which inference did Darwin make about the connection between eye size and living underground?

   (1) Animals with large eyes are unable to survive underground.
   (2) The eyes of burrowing animals are smaller, but the animals' eyesight is stronger.
   (3) Burrowing in the ground can cause an animal to develop larger eyes.
   (4) The small eyes of burrowing animals are beneficial to survival.
   (5) The eyes of moles are smaller than eyes of other burrowing rodents.

③ Apply the Skill

Directions: Choose the one best answer to each question.

Questions 2, 3, and 4 refer to the paragraph and illustration.

Charles Darwin left England in 1831 for a voyage around the world. The trip gave him the chance to observe and collect plants and animals from many different places. In late 1835, Darwin spent several weeks on the Galápagos Islands off the coast of South America. He collected several finches from the islands and brought the specimens back to London. Once home, Darwin studied the birds. He was amazed to find that, although they were all finches from the same group of islands, they had beaks of different sizes and shapes. Darwin wrote that, "One might really fancy that . . . one species had been taken and modified for different ends."

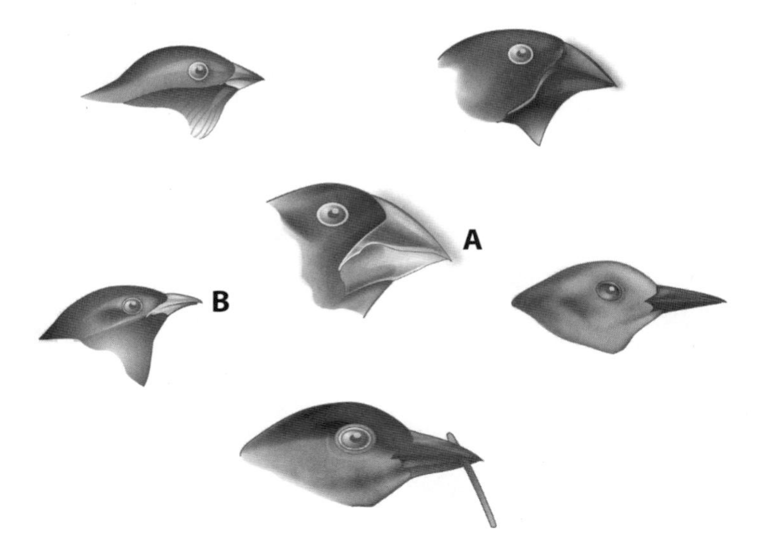

2. In the passage, Darwin states that one might think one species had been modified for different purposes. What inference did Darwin make?

   (1) The different beaks prove that all birds have different purposes.
   (2) Finches from different islands had different beak shapes.
   (3) Birds on the Galápagos are different from birds in London.
   (4) The different finches had come from the same ancestor.
   (5) Climate affects the shape of bird beaks.

3. On what did Darwin base his inference?

   Darwin based his inference on

   (1) personal experience
   (2) the work of another scientist
   (3) his own observations
   (4) a prediction about life on the Galápagos
   (5) similar finches in London

4. Based on the information in the illustration, what inference can you make about Bird A and Bird B?

   (1) Bird A is more likely than Bird B to survive in its environment.
   (2) Bird B is an ancestor of Bird A.
   (3) Bird B's beak is thicker than Bird A's beak.
   (4) The birds probably do not have to compete for food.
   (5) The birds are found on the same island.

Question 5 refers to the paragraph.

As a young scientist, Charles Darwin spent five years observing nature while on board the *Beagle*. While exploring parts of South America, Darwin experienced first-hand the devastation caused by a volcanic eruption and by an earthquake. The earthquake, in Chile, led to a tidal wave that destroyed the Chilean city of Concepción. Darwin walked through the rubble of the city and was intrigued both by the damage as well as by a strange sight: Local mussel beds, all dead, were now above high tide.

5. What did Darwin infer about the surrounding land from his observation of mussel beds above high tide?

   (1) The land had risen.
   (2) The land had fallen.
   (3) The earthquake had replenished the mussel population.
   (4) The eruption had occurred after the earthquake.
   (5) The mussel beds had been buried beneath the city of Concepción.

LESSON 15

Draw Conclusions

1 Learn the Skill

Remember that an inference is an educated guess based upon facts, evidence, experience, or reasoning. A **conclusion** is an explanation or judgment that generally relies on inferences. When you **draw conclusions**, you make a statement that explains your observations and the facts that you have.

2 Practice the Skill

By mastering the skill of drawing conclusions, you will improve your study and test-taking skills, especially as they relate to the GED Science Test. Look at the diagram and strategies below. Then answer the question that follows.

A A conclusion is the answer to a researcher's main question. In this case, Mendel carried out his experiments to determine how traits are passed from one generation to the next.

B A conclusion differs from a summary. For example, the statement, "Some offspring produced purple flowers and some produced white flowers" is a summary of the observations, not a conclusion. A conclusion is an explanation or judgment that is based on an educated guess.

Gregor Mendel's experiments with pea plants in the mid-1800s laid the foundation for the science of genetics. <u>Mendel wanted to learn how traits were passed down and combined through generations.</u> The diagram shows the results of an experiment in which Mendel bred plants with purple flowers to plants with white flowers.

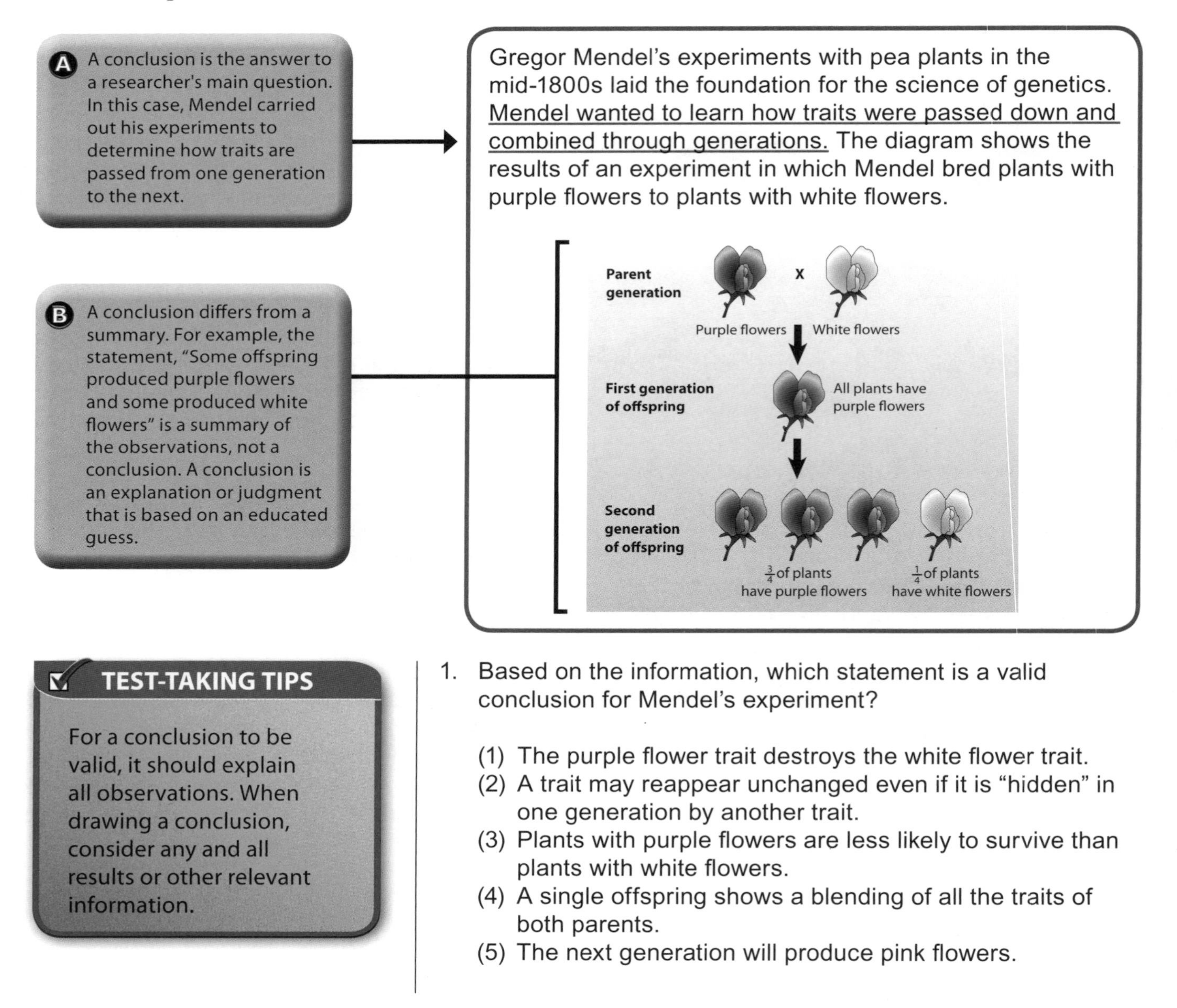

TEST-TAKING TIPS

For a conclusion to be valid, it should explain all observations. When drawing a conclusion, consider any and all results or other relevant information.

1. Based on the information, which statement is a valid conclusion for Mendel's experiment?

   (1) The purple flower trait destroys the white flower trait.
   (2) A trait may reappear unchanged even if it is "hidden" in one generation by another trait.
   (3) Plants with purple flowers are less likely to survive than plants with white flowers.
   (4) A single offspring shows a blending of all the traits of both parents.
   (5) The next generation will produce pink flowers.

3 Apply the Skill

Directions: Choose the one best answer to each question.

Questions 2, 3, and 4 refer to the following paragraph and diagram.

Mendel drew certain conclusions from the results of his experiments with pea plants. He realized that each plant carried pairs of "heritable factors" (what we now call *genes*) for all its visible traits. Mendel's results demonstrated that certain inherited traits show dominance or recessiveness. The Punnett square below represents the cross between two purple flowers, both of which had one copy of the white-flower trait and one copy of the purple-flower trait. The boxes inside the square show the possible combinations of flower-color traits in the offspring.

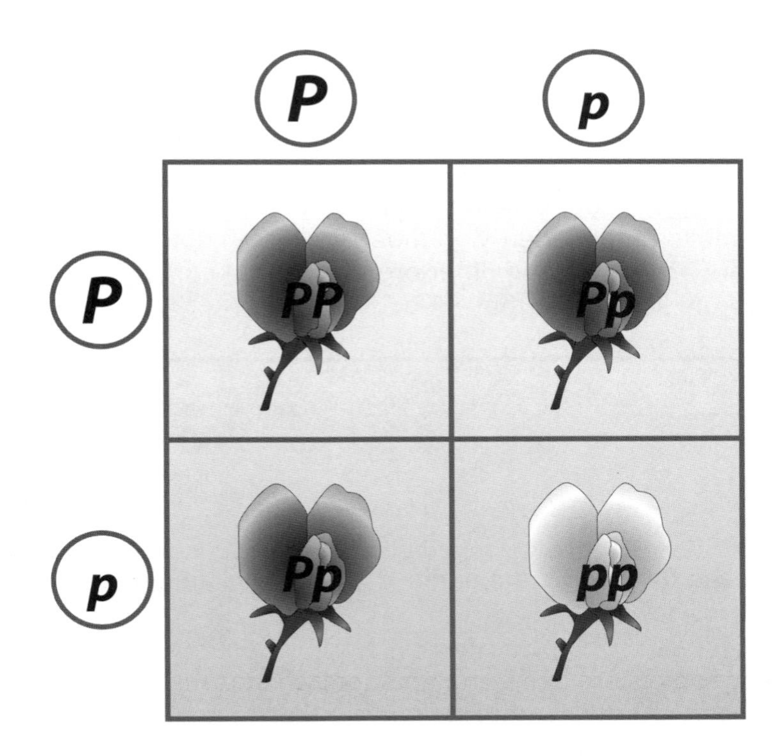

2. Based on the information, what conclusion can you make about the relationship between the white-flower trait and the purple-flower trait?

   The white-flower trait

   (1) shows even when the individual also has a copy of the purple-flower trait
   (2) never appears in the offspring of two purple-flowered plants
   (3) is dominant to the purple-flower trait
   (4) is recessive to the purple-flower trait
   (5) is visible in all pea plants

3. Based on the information in the diagram, what can you conclude about why the color of the pea flowers in each generation alone does not predict the flower color of the next generation?

   (1) Color changes randomly with each generation.
   (2) An individual showing the dominant trait may be able to pass on a copy of the recessive trait to its offspring.
   (3) Plants generally lack sufficient offspring for a recessive trait to appear.
   (4) The trait for white flower color disappears after the first generation.
   (5) An individual showing the recessive trait may carry a hidden version of the dominant trait.

4. Based on the information, what conclusion can you make about the parents of all pea plants with white flowers?

   The parents of all pea plants with white flowers

   (1) have white flowers
   (2) have at least one copy apiece of the white-flower trait
   (3) produce only offspring with white flowers
   (4) produce only offspring with purple flowers
   (5) each have two copies of the white-flower trait

LESSON 16

Generalize

1 Learn the Skill

When you **generalize**, you make a broad statement that applies to an entire group of people, places, or events. A generalization can be valid or invalid. **Valid generalizations** are conclusions supported by facts and examples. **Invalid generalizations** are conclusions that are unsupported by facts or examples.

2 Practice the Skill

By mastering the skill of generalizing, you will improve your study and test-taking skills, especially as they relate to the GED Science Test. Read the paragraph and strategies below. Then answer the question that follows.

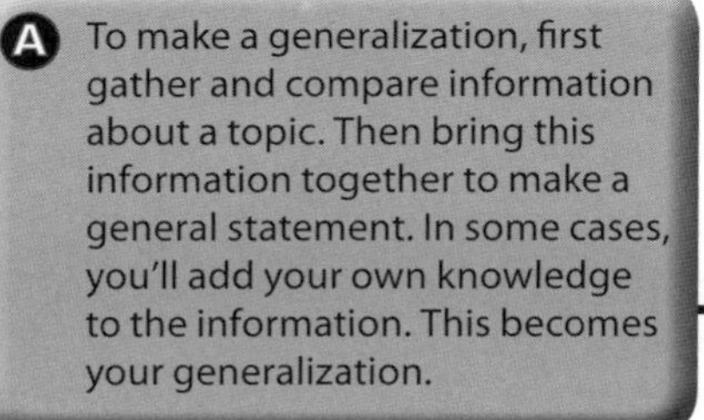

B Even if a fact or definition describes all the members of a particular group, it is not a generalization. In this case, the fact that producers make their own food is their defining characteristic. So this statement is not a *generalization* about producers. It is a *fact* about producers.

An ecosystem is made up of a community of living organisms and the nonliving parts of the environment. Energy flows through the living components of an ecosystem. An organism gets energy from its food and passes on energy to any organism that feeds on it. Living things in an ecosystem can be classified into two broad groups based on how they obtain energy. (B) Producers can make their own food. Consumers must feed on other organisms to obtain energy. In most ecosystems, the producers are plants and other organisms that can use energy from sunlight to make food.

TEST-TAKING TIPS

When you make a generalization, consider several examples of people, places, or things. Then consider what they have in common. Generalizations may contain key words such as *all, always, every,* and *never,* as well as *most, mostly, typically, in general, generally, often, overall, almost,* and *usually.*

1. Which of the following statements is a generalization made by the passage?

    (1) Every ecosystem has living and nonliving parts.
    (2) Most producers use energy from sunlight to make food.
    (3) Energy flows between the organisms in an ecosystem.
    (4) Consumers are organisms that feed on other living things.
    (5) Organisms in an ecosystem can be divided into two main groups: producers and consumers.

③ Apply the Skill

Directions: Choose the one best answer to each question.

Questions 2 and 3 refer to the following paragraph.

An ecosystem contains different kinds of living things. Scientists often think of the living things in an ecosystem as being organized into different levels. The lowest level of organization is the individual organism. Individual organisms are organized into populations. A population is a group of organisms that are all the same species and that all live in the same area. For example, all of the blue jays in a forest make up a population. Populations are organized into communities. A community is all of the populations in an area. Most communities contain many different populations.

2. Based on the information in the paragraph, which statement is a valid generalization about the living things in an ecosystem?

   (1) There are different kinds of living things in an ecosystem.
   (2) The communities in most ecosystems are made up of many different species.
   (3) Populations in an ecosystem are made up of individual organisms.
   (4) All of the organisms in a population are of the same species.
   (5) The individual organisms in a community are of many different species.

3. Which statement is an invalid generalization about the living things in an ecosystem?

   (1) A population includes only organisms from the same species that live in the same area.
   (2) Each community includes the same populations.
   (3) Communities contain individual organisms from different areas.
   (4) Organisms are organized into populations, which in turn make up communities.
   (5) Scientists often group living things into different levels.

Questions 4 and 5 refer to the following paragraph.

Nonliving things can be very important in an ecosystem. For example, all living things need water to survive, but different living things need different amounts of water. So the amount of rainfall that an ecosystem receives helps to determine which organisms can live within it. In addition, an ecosystem's amount of sunlight and temperature similarly affect the organisms that can live there. Most organisms survive best within a narrow range of conditions. Only a few kinds of organisms can survive in very hot, very cold, very dry, or very dark environments.

4. Which statement is a valid generalization based on information in the paragraph?

   (1) Most organisms live in ecosystems with mild temperatures.
   (2) The temperature range in an ecosystem affects which organisms live in the ecosystem.
   (3) All living things need water to survive.
   (4) Organisms require different amounts of sunlight to survive.
   (5) Nonliving things can affect the organisms in an ecosystem.

5. Which statement is a valid generalization made in the paragraph?

   (1) Living things depend on nonliving things to survive.
   (2) Some organisms need more water than others.
   (3) Most organisms need specific environmental conditions to survive.
   (4) The amount of sunlight in an ecosystem affects the organisms that live there.
   (5) Organisms can live in all types of conditions.

UNIT 1

LESSON 17

Questioning

1 Learn the Skill

Questioning is the process in which you ask a question about something that you observe. Questioning helps you review and understand what you read. It also helps you to identify any gaps in knowledge about a subject and then find ways to fill in those gaps with missing information.

2 Practice the Skill

By mastering the skill of questioning, you will improve your study and test-taking skills, especially as they relate to the GED Science Test. Look at the text, illustration, and strategies below. Then answer the question that follows.

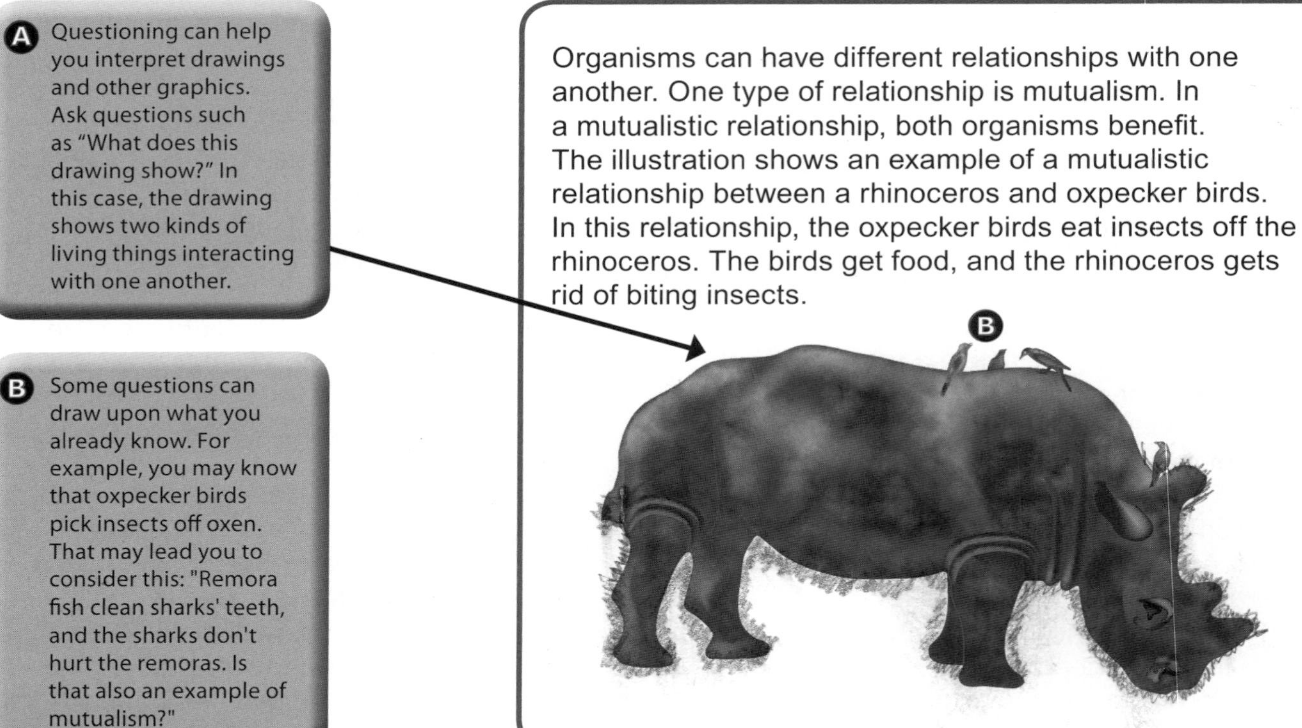

A Questioning can help you interpret drawings and other graphics. Ask questions such as "What does this drawing show?" In this case, the drawing shows two kinds of living things interacting with one another.

B Some questions can draw upon what you already know. For example, you may know that oxpecker birds pick insects off oxen. That may lead you to consider this: "Remora fish clean sharks' teeth, and the sharks don't hurt the remoras. Is that also an example of mutualism?"

Organisms can have different relationships with one another. One type of relationship is mutualism. In a mutualistic relationship, both organisms benefit. The illustration shows an example of a mutualistic relationship between a rhinoceros and oxpecker birds. In this relationship, the oxpecker birds eat insects off the rhinoceros. The birds get food, and the rhinoceros gets rid of biting insects.

USING LOGIC

Asking questions about the material, such as "What do I already know about this subject?" can help you make connections to your prior knowledge.

1. Which question could you ask to help you understand why the illustration is an example of a mutualistic relationship?

   (1) How large are the oxpecker birds?
   (2) How many insects does an oxpecker bird eat in a day?
   (3) Where do the oxpecker birds and rhinoceros live?
   (4) How do the oxpecker birds and rhinoceros help each other?
   (5) What does a rhinoceros eat?

3 Apply the Skill

Directions: Choose the one best answer to each question.

Questions 2 and 3 refer to the following paragraph.

Symbiosis is a situation in which two organisms from different species live closely together. There are several types of symbiotic relationships. One type is commensalism. In a commensal relationship, one organism benefits. The other organism does not benefit, but it is not harmed, either. For example, some types of orchids have a commensal relationship with the trees on which they grow. These orchids grow on the branches of large trees in tropical forests. The trees receive no benefit, but the orchids do not harm them.

2. Which question should you ask to help you understand why the orchid-tree relationship is a commensal relationship?

   (1) How large are the orchids?
   (2) How tall do the trees grow?
   (3) How do the orchids benefit from the relationship?
   (4) How are the orchids attached to the tree?
   (5) In which type of soil do the trees and orchids grow?

3. What question could you ask to determine whether the orchid-tree relationship is also an example of mutualism?

   (1) Does either organism benefit from the relationship?
   (2) How do the trees benefit from the relationship?
   (3) How would a more moderate climate affect the relationship between trees and orchids?
   (4) How are the trees harmed by the orchids?
   (5) How does the size of the tree branches affect the orchids?

Questions 4 and 5 refer to the following paragraph and the graph.

Some animals have a predator-prey relationship. A predator is an animal that catches and eats organisms of another species. The organisms that it catches and eats are its prey. The graph below shows the relationship between the populations of lynxes and hares in an ecosystem over time. In this ecosystem, the lynx is a predator, and the hare is its prey.

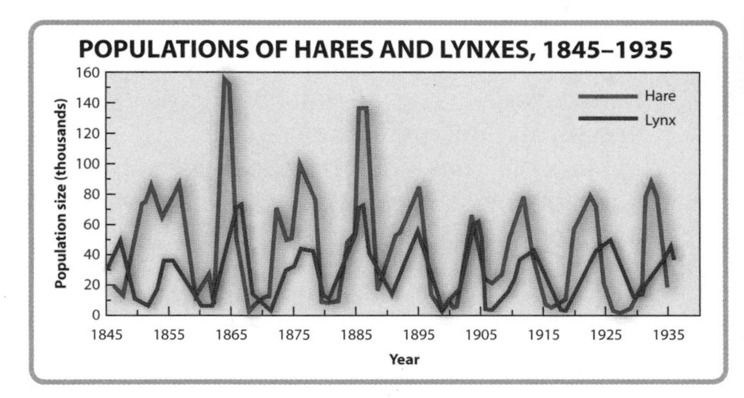

4. Which question could the graph help you answer?

   (1) What are characteristics that make lynxes good predators?
   (2) What happens to the number of lynxes when the number of hares increases?
   (3) How do hares hide from lynxes?
   (4) How do lynxes and hares survive cold weather?
   (5) Are hares predators for any organisms?

5. If you wanted to understand more about predator-prey relationships, which question would it be most useful for you to ask?

   (1) Is the lynx-hare population graph similar to graphs of other predator-prey populations?
   (2) Where do lynxes and hares live?
   (3) How many hares can a lynx kill in its lifetime?
   (4) Why were there so many hares in 1865?
   (5) Are hares prey for any other predators?

Unit 1 Review

The Unit Review is structured to resemble the GED Science Test. Be sure to read each question and all possible answers very carefully before choosing your answer.

To record your answers, fill in the numbered circle that corresponds to the answer you select for each question in the Unit Review.

Do not rest your pencil on the answer area while considering your answer. Make no stray or unnecessary marks. If you change an answer, erase your first mark completely.

Mark only one answer space for each question; multiple answers will be scored as incorrect.

Sample Question

How does a diagram differ from other forms of graphics?

A diagram

(1) organizes information in rows and columns
(2) shows relationships visually among objects or events
(3) summarizes information in a passage
(4) shows parts of a whole
(5) shows change over time

Directions: Choose the one best answer to each question.

Questions 1 and 2 refer to the following graph.

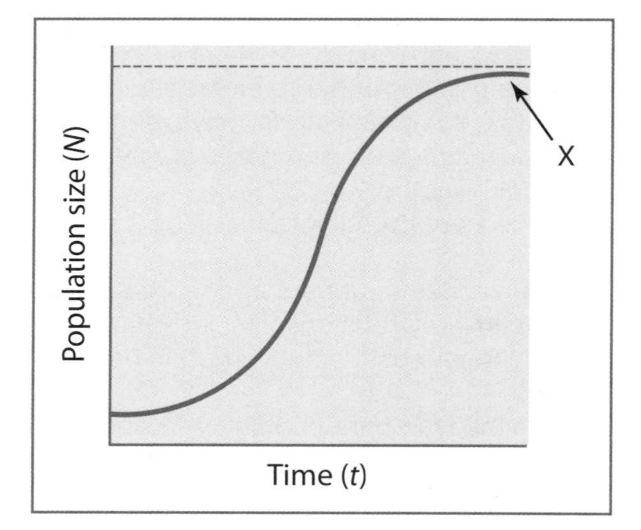

1. The graph shows the growth of one population in a grassland ecosystem. What happened to the population at point *X*?

   At point *X*, the population

   (1) began to grow more quickly
   (2) suddenly decreased
   (3) stopped growing
   (4) began to grow more slowly
   (5) disappeared from the ecosystem

   ①②③④⑤

2. Based on the graph, which of the following is a logical cause for what happened to the population at point *X*?

   A logical cause of the trend shown at point *X* is

   (1) a limited food supply
   (2) an introduced predator
   (3) a sudden disease
   (4) availability of unlimited resources
   (5) inability of individuals to find mates

   ①②③④⑤

Question 3 refers to the paragraph.

Plants adapt to the environment in which they live. Many desert plants have stems that store water and leaves with a waxy coating to decrease water loss. Thin, wax-coated needles on pine trees limit heat and moisture loss in the cold of winter.

3. Which statement best summarizes the paragraph above?

   (1) Plants have adaptations that allow them to live in different environments.
   (2) Desert plants have adaptations for surviving in hot climates.
   (3) The thin needles of pine trees limit heat loss.
   (4) Desert plants often have stems that help them retain scarce water.
   (5) Pine tree adaptations allow them to survive in cold climates.

①②③④⑤

4. The illustration below represents the nitrogen cycle.

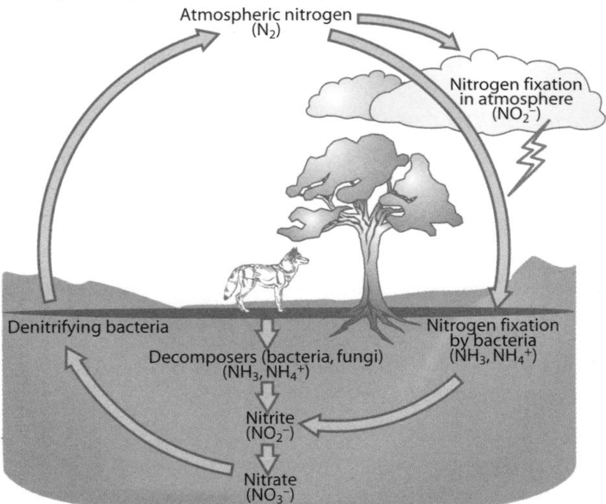

Based on the illustration, which statement correctly describes an event of the nitrogen cycle?

   (1) Plants take N_2 directly from the air.
   (2) Plants release N_2 to the atmosphere, where it is converted to nitrites.
   (3) Some bacteria in the soil recycle nitrogen back into the atmosphere.
   (4) Lightning adds N_2 to the atmosphere.
   (5) The supply of nitrogen on Earth decreases each time nitrogen fixation happens.

①②③④⑤

Questions 5 and 6 refer to the following information and diagram.

In pea plants, purple flower color is a dominant trait and white flower color is a recessive trait. This means that white flower color will appear only if an individual offspring receives the recessive trait from both parents. *P* represents the dominant trait, and *p* represents the recessive trait. A cross between two particular pea plants resulted in an individual offspring that produced white flowers. The Punnett square below shows all the possible flower colors for other offspring from this cross.

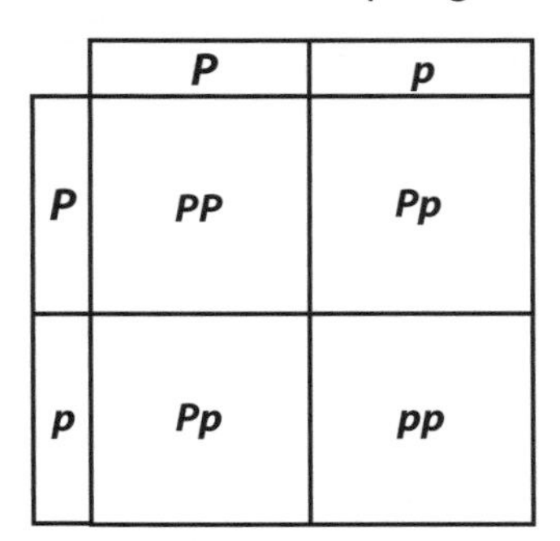

	P	**p**
P	**PP**	**Pp**
p	**Pp**	**pp**

5. What conclusion can you draw about the parent plants?

   (1) Both parents carried a trait for pink flowers.
   (2) One parent carried the dominant trait for purple flowers, and the other parent carried the recessive trait for white flowers.
   (3) Each parent had at least one recessive trait for white flowers.
   (4) Both parents could produce both purple and white flowers.
   (5) One parent experienced a mutation in the trait for flower color.

①②③④⑤

6. Based on the diagram, what percent of total offspring of this cross will most likely have purple flowers?

   (1) 20%
   (2) 25%
   (3) 50%
   (4) 75%
   (5) 100%

①②③④⑤

UNIT 1

UNIT 1

Questions 7 through 9 refer to the following paragraph and diagram.

Sea otters live in kelp forests along the Pacific coast of North America. Biologists found that when the population of sea otters decreases, the population of sea urchins increases. The urchins then deplete the kelp populations.

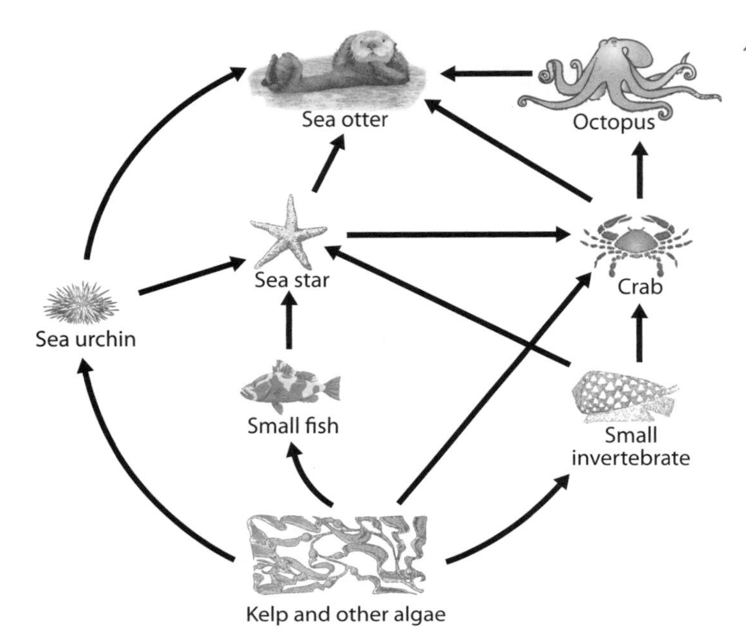

7. Which of these organisms in a kelp forest ecosystem is a producer?

(1) sea otter
(2) kelp
(3) sea urchin
(4) sea star
(5) crab

①②③④⑤

8. Which organism does **NOT** feed on kelp and other algae?

(1) sea urchin
(2) small invertebrate
(3) small fish
(4) crab
(5) sea star

①②③④⑤

9. Based on the information and diagram, which statement about organisms in a kelp forest is most accurate?

(1) The otter provides food for all other species of animals in this ecosystem.
(2) Sea otters are the most important producers in this ecosystem.
(3) The sea otter is a marine mammal.
(4) Without the otter, the populations of many other species in this ecosystem would increase.
(5) All of the other plants and animals in this ecosystem would disappear without the sea otter.

①②③④⑤

Question 10 refers to the following excerpt about the Galápagos Islands from Charles Darwin.

"Considering that these islands are placed directly under the equator, the climate is far from being excessively hot, but this seems chiefly to be caused by the singularly low temperature of the surrounding water, brought here by the great southern Polar current . . ."

10. Based on the excerpt, what did Darwin infer about the climate of the Galápagos Islands?

Darwin inferred that

(1) the climate was hot because the islands are near the equator
(2) the relatively cool climate was due to cool waters around the islands
(3) the islands have a polar climate because they are near the polar current
(4) the surrounding seas make the climate wetter than many other places
(5) the islands have fairly low temperatures because they are below the equator

①②③④⑤

Questions 11 and 12 refer to the following timeline.

The timeline below shows major events in the history of red wolf conservation in the United States.

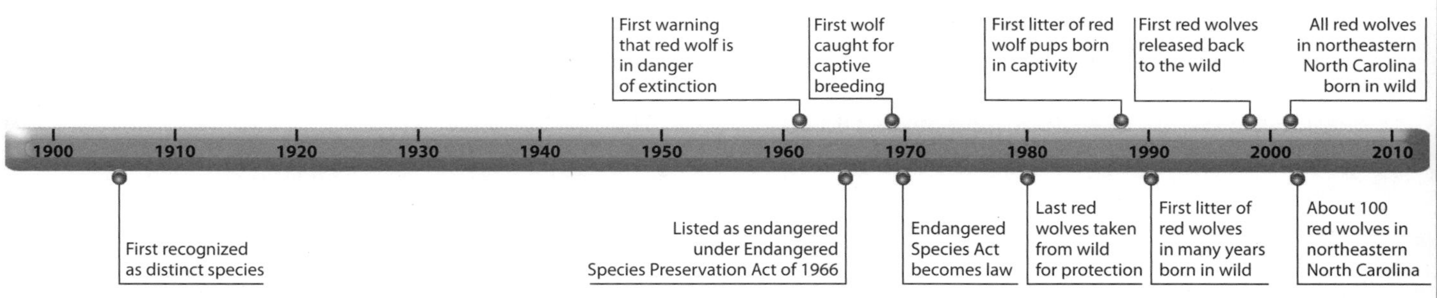

11. Based on the timeline, which of the following statements is true?

    (1) There have been no red wolves in the wild since 1980.
    (2) About eight years passed before captive red wolves were bred successfully.
    (3) Red wolves became extinct in the wild in 1962.
    (4) Captive breeding of wolves made the Endangered Species Act unnecessary.
    (5) The Endangered Species Preservation Act replaced the Endangered Species Act.

①②③④⑤

12. Based on the timeline, which statement summarizes the history of red wolf conservation in the United States?

    (1) The Endangered Species Act was the starting point for efforts to protect red wolves.
    (2) A combination of laws and captive breeding has had a positive effect on red wolf populations.
    (3) Red wolves remain entirely in captivity today.
    (4) Red wolves were recognized as a separate wolf species in the early 1900s.
    (5) Red wolves became endangered due to the Endangered Species Protection Act.

①②③④⑤

13. Carnivores are organisms that feed only on animals. Herbivores are organisms that feed only on plants. Omnivores feed on plants and animals.

EXAMPLES OF ANIMAL SPECIES AND THEIR FEEDING ROLES IN ECOSYSTEMS

SPECIES (COMMON NAME)	FEEDING ROLE
American alligator	Carnivore
Lion	Carnivore
Black bear	Omnivore
Rat snake	Carnivore

Based on the information and table, what inference can you make about the alligator?

The alligator

    (1) is found in the same ecosystem as the rat snake
    (2) feeds only on herbivores
    (3) is not prey for any other organism
    (4) is higher on the food chain than the lion
    (5) may feed on herbivores, carnivores, and omnivores

①②③④⑤

UNIT 1

Questions 14 and 15 refer to the following information.

A virus is basically a bundle of genetic material surrounded by a protein shell. Viruses are tiny—even smaller than bacteria. Scientists classify viruses as microbes, but viruses are not living things like other microbes. Unlike living things, viruses cannot reproduce themselves. Instead, they need living cells to reproduce. When a virus enters a living cell, it takes over the cell's processes. The virus directs the cell to copy the genetic material of the virus, making new virus particles. The new viruses kill the host cell. Then they spread to neighboring cells. If nothing stops them, the viruses keep reproducing and spreading through living cells.

14. Which statement best expresses the main idea of the passage?

(1) Viruses reproduce inside living cells.
(2) Viruses are extremely small.
(3) Viruses kill their host cells after reproducing inside them.
(4) Viruses are made of both protein and genetic material.
(5) Viruses are not alive.

①②③④⑤

15. Which of these details supports the main idea of the passage?

Viruses

(1) are smaller than bacteria
(2) are surrounded by a protein shell
(3) resemble microbes
(4) contain genetic material
(5) take over a cell's processes to reproduce themselves

①②③④⑤

Questions 16 and 17 refer to the following table.

The table below identifies organisms found in a freshwater ecosystem.

SPECIES	PRIMARY DIET
Clamworm	Zooplankton
Cordgrass	N/A
Herring gull	Soft-shelled clam, smelt
Marsh periwinkle	Cordgrass
Peregrine falcon	Herring gull, snowy egret, short-billed dowitcher
Phytoplankton	N/A
Short-billed dowitcher	Clamworm, marsh periwinkle
Snowy egret	Smelt
Soft-shelled clam	Phytoplankton
Smelt	Zooplankton
Zooplankton	Phytoplankton

16. Based on the table, which of these simple food chains is most likely?

(1) herring gull → clam → phytoplankton
(2) cordgrass → marsh periwinkle → short-billed dowitcher
(3) short-billed dowitcher → cordgrass → peregrine falcon
(4) peregrine falcon → herring gull → cordgrass
(5) clamworm → zooplankton → cordgrass

①②③④⑤

17. Based on the information, what would be the most likely effect of removing cordgrass from the ecosystem?

(1) The populations of marsh periwinkles and short-billed dowitchers would decrease.
(2) All consumers would become extinct.
(3) Phytoplankton populations would increase.
(4) Peregrine falcons would have greater food resources.
(5) Marsh periwinkles would begin to feed on zooplankton.

①②③④⑤

18. Biologists interested in protecting polar bear populations have fitted individual bears with radio collars and tracked the individuals throughout several years. The radio collars give scientists information about a bear's location and movements, and can allow scientists to determine whether a particular bear is still alive.

Which of the following questions could be answered with the wildlife study method described above?

(1) How many polar bears exist in the wild?
(2) What is the average survival rate of adult polar bears over a five-year period?
(3) What is the preferred food of polar bears?
(4) How many cubs does a female polar bear have each year?
(5) How much weight does a polar bear gain during seal-hunting season?

①②③④⑤

19. Acid deposition can affect the growth of plants. Over time, acid deposition raises the acid level of soil. The acid can cause water in the soil to leach important nutrients, such as magnesium and calcium. Plants do not grow as well when the soil does not contain a sufficient supply of these nutrients.

Based on the information in the passage, what is the meaning of *leach*?

To *leach* is to

(1) poison
(2) accumulate
(3) make stronger
(4) remove
(5) produce

①②③④⑤

20. The diagram below shows the human digestive system.

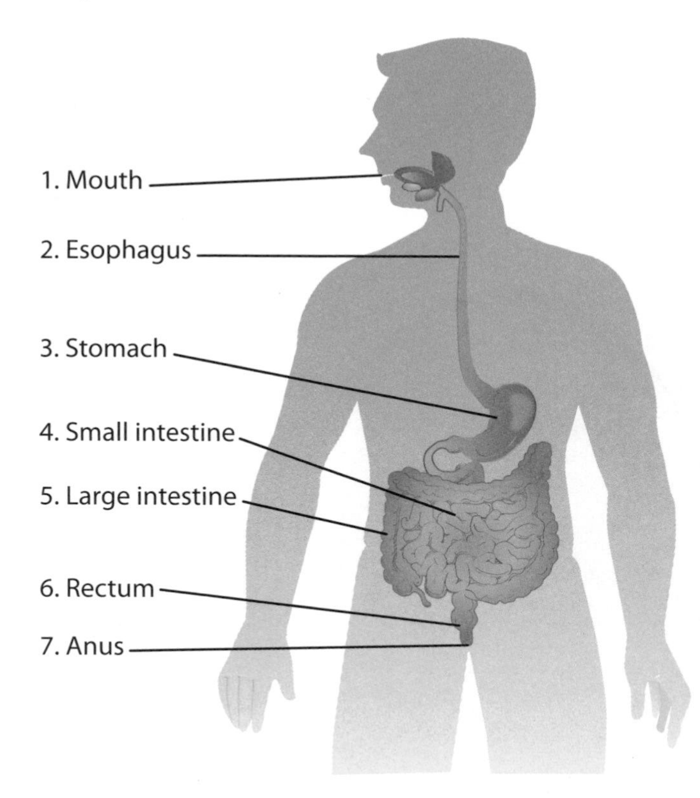

Based on the diagram, what must happen before food mixes with digestive juices in the stomach?

(1) Waste must accumulate in the large intestine.
(2) The body must absorb nutrients from the food.
(3) Waste must pass out of the rectum.
(4) The food must travel down the esophagus.
(5) The pancreas must send digestive juices into the small intestine.

①②③④⑤

UNIT 1

GED JOURNEYS

F. STORY MUSGRAVE

F. Story Musgrave's life—and career—took off after he earned his GED certificate. Among other achievements, Musgrave became the first astronaut to fly missions on all five space shuttles.

F. Story Musgrave wanted to reach for the stars. So in 1953, Musgrave put his education on hold, leaving St. Mark's School in Massachusetts to join the military. As a member of the U.S. Marine Crops, Musgrave served as an aviation technician and aircraft crew chief and earned his GED certificate while stationed in Hawaii. It served as a springboard for his many successes to come.

After completing his military commitment, Musgrave earned the first of many college degrees, this one in mathematics and statistics from Syracuse (N.Y.) University in 1958. He went on to earn another undergraduate degree (in chemistry, 1959) and a trio of master's degrees, all in different areas of study (business administration, UCLA; physiology and biophysics, University of Kentucky; literature, University of Houston). He also earned a doctorate in medicine from Columbia (N.Y.) University.

But in many ways, Musgrave's career had only begun to take flight. In 1967, he was selected by NASA as a scientist-astronaut. Over time, Musgrave participated in the design and development of spacesuits, life support systems, airlocks, and other features related to the shuttle program. Musgrave first reached space in 1983 as a crew member on the shuttle *Challenger.* He later became the only astronaut to fly missions on all five space shuttles. Musgrave performed numerous mission functions, including several famous spacewalks. He earned the nickname "Dr. Detail" for his disciplined efforts. As he noted,

> **"You want to get the mission done, so you aim at perfection. You aim at perfecting your art of working in space."**

All told, Musgrave flew 17,700 hours in 160 different types of civilian and military aircraft. An accomplished parachutist, he made more than 500 jumps. He retired from NASA in 1997 after 30 years of service.

BIO BLAST: F. Story Musgrave

- Born in 1935 in Boston, Massachusetts
- Served with U.S. Marines in Korea, Japan, and Hawaii
- Earned a total of six college degrees
- Flew on the maiden voyage of shuttle *Challenger*
- Performed three spacewalks during the *Endeavour* mission to repair the Hubble Space Telescope

Unit 2: Earth/Space Science

Every day, whether or not you realize it, you study Earth science. Weather forecasts, reports on climate change, stories about renewable energy—all these provide information about the world around us, how the world affects us, and how we affect that world.

Earth and space science also plays an important part in the GED Science Test, comprising 20 percent of the questions. As with other areas of the GED Science Test, Earth and space science questions will test your ability to read and analyze different types of text and graphics. Unit 2 spotlights the introduction and interpretation of various diagrams, maps, flowcharts, and other core science concepts that will help you prepare for the GED Science Test.

Table of Contents

Lesson 1: Understand and Evaluate a Hypothesis. 44–45
Lesson 2: Interpret 3-Dimensional Diagrams 46–47
Lesson 3: Understand Maps and Map Symbols. 48–49
Lesson 4: Interpret Physical and Topographic Maps. 50–51
Lesson 5: Interpret Flowcharts 52–53
Lesson 6: Compare and Contrast Visuals 54–55
Lesson 7: Determine Fact and Opinion. 56–57
Lesson 8: Evaluate Information 58–59
Lesson 9: Identify Problem and Solution. 60–61
Unit 2 Review 62–67

LESSON 1

Understand and Evaluate a Hypothesis

1 Learn the Skill

In everyday speech, people often use the terms *hypothesis* and *theory* to mean "a guess." A **hypothesis** is a guess, but one based on evidence and logic. In science, however, a theory is definitely not a guess. A theory is an explanation that is supported by all the available data. To **understand and evaluate a hypothesis**, you need to make sure there is some evidence to support the guess. You will also need to be sure that a hypothesis can be restated in the form of a question that could be tested by a scientific investigation.

2 Practice the Skill

By mastering the skill of understanding and evaluating a hypothesis, you will improve your study and test-taking skills, especially as they relate to the GED Science Test. Read the paragraph and strategies below. Then answer the question that follows.

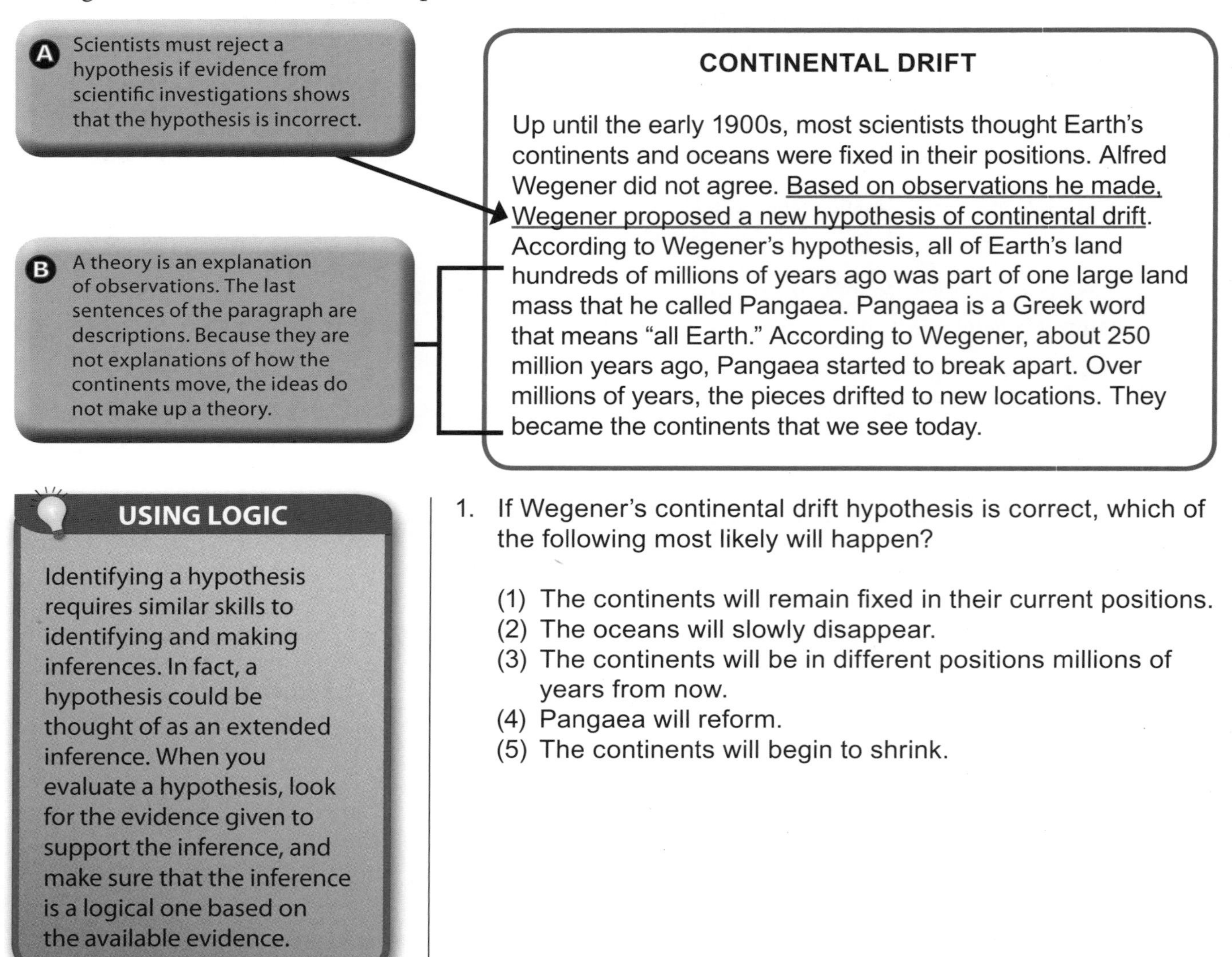

A Scientists must reject a hypothesis if evidence from scientific investigations shows that the hypothesis is incorrect.

B A theory is an explanation of observations. The last sentences of the paragraph are descriptions. Because they are not explanations of how the continents move, the ideas do not make up a theory.

CONTINENTAL DRIFT

Up until the early 1900s, most scientists thought Earth's continents and oceans were fixed in their positions. Alfred Wegener did not agree. Based on observations he made, Wegener proposed a new hypothesis of continental drift. According to Wegener's hypothesis, all of Earth's land hundreds of millions of years ago was part of one large land mass that he called Pangaea. Pangaea is a Greek word that means "all Earth." According to Wegener, about 250 million years ago, Pangaea started to break apart. Over millions of years, the pieces drifted to new locations. They became the continents that we see today.

USING LOGIC

Identifying a hypothesis requires similar skills to identifying and making inferences. In fact, a hypothesis could be thought of as an extended inference. When you evaluate a hypothesis, look for the evidence given to support the inference, and make sure that the inference is a logical one based on the available evidence.

1. If Wegener's continental drift hypothesis is correct, which of the following most likely will happen?

   (1) The continents will remain fixed in their current positions.
   (2) The oceans will slowly disappear.
   (3) The continents will be in different positions millions of years from now.
   (4) Pangaea will reform.
   (5) The continents will begin to shrink.

③ Apply the Skill

Directions: Choose the one best answer to each question.

Questions 2 and 3 refer to the following paragraph, excerpt, and drawing.

Wegener based his ideas about Earth's continents on several pieces of evidence. One piece of evidence involved the shapes of South America and Africa. Wegener wrote:

"It is just as if we were to refit the torn pieces of a newspaper by matching their edges and then check whether the lines of print ran smoothly across. If they do, there is nothing left but to conclude that the pieces were in fact joined in this way."
From Wegener's *The Origin of Oceans and Continents*

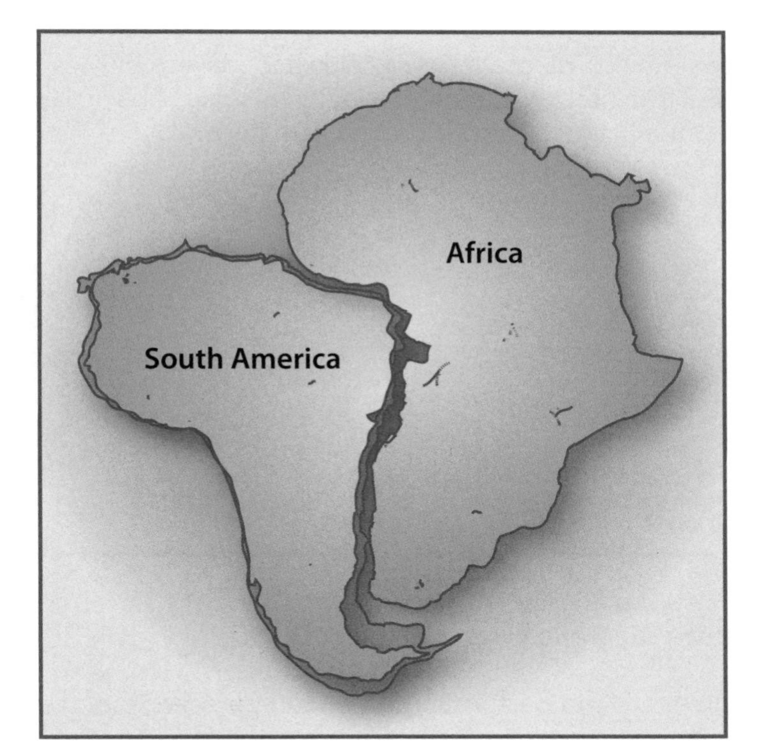

2. Which statement best explains the importance of Wegener's observation about the "fit" of the coasts of South America and Africa?

   (1) It provided evidence for the idea that continents move.
   (2) It showed how the continents moved to their current positions.
   (3) It was based on reasoning instead of data.
   (4) It did not need to be tested.
   (5) It proved that Africa and South America share many similarities today.

3. Which additional piece of evidence would support Wegener's hypothesis that Africa and South America once had been joined?

   (1) different plant and animals species found on the two continents
   (2) a lack of fossils on both continents
   (3) the presence of similar fossils on both continents
   (4) similar climates today on both continents
   (5) a map drawn by another scientist showing how the continents might have looked

Question 4 refers to the following paragraph.

Until the 1960s, scientists still did not know the mechanism that allows continents to "drift," or move. With the development of new technologies and the collection of more data, scientists got a clearer picture not only of how continents move but of how this movement is related to other Earth processes and features. According to the theory of plate tectonics, Earth's crust is broken up into several large plates as well as many smaller ones, and the plates move on a solid (but flowing) Earth layer. This theory explains not only how continents moved to their present locations but also the formation of land features, such as mountains, and geological events, such as volcanic eruptions.

4. Which statement best explains why plate tectonics is a theory and not a hypothesis?

   Plate tectonics

   (1) explains many observations
   (2) is a guess about how the continents moved
   (3) developed from the continental drift hypothesis
   (4) can never be disproven
   (5) was developed based on evidence

LESSON 2

Interpret 3-Dimensional Diagrams

1 Learn the Skill

Many **3-dimensional (3-D) diagrams** show part of an object or structure cut away so that the inside of the object is visible. Such cutout diagrams are most often used to show the layers that make up a structure or object. To **interpret 3-dimensional diagrams**, pay attention to how the outside of the object is related to the inside features shown in the diagram.

2 Practice the Skill

By mastering the skill of interpreting 3-dimensional diagrams, you will improve your study and test-taking skills, especially as they relate to the GED Science Test. Examine the diagram and callouts and read the strategies below. Then answer the question that follows.

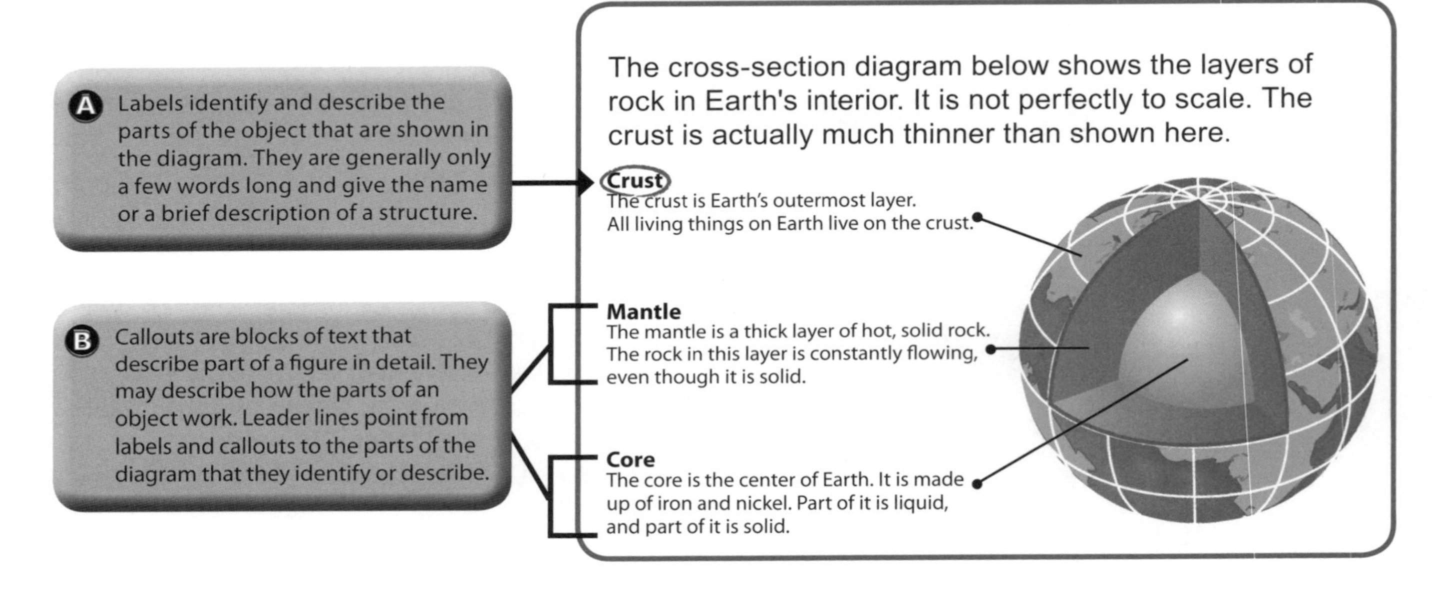

MAKING ASSUMPTIONS

Although a 3-D diagram represents what the actual object looks like, a 3-D diagram is not always to scale. A scale diagram is one in which the relative sizes of the parts of the diagram are the same as they are in real life. Unless a diagram specifically states that it is drawn to scale, assume that it is not to scale.

1. Based on the diagram, what can you conclude?

   (1) The mantle is the part of Earth that can be seen from the surface.
   (2) The core is less dense than the mantle.
   (3) Temperature inside Earth increases with depth.
   (4) All of Earth's layers are equally thick.
   (5) The crust makes up only a very small portion of Earth.

3 Apply the Skill

Directions: Choose the one best answer to each question.

Questions 2 and 3 refer to the following passage and diagram.

Earth's crust is broken up into several large pieces called tectonic plates. The tectonic plates move slowly—about as fast as fingernails grow. As the plates move, they collide, pull apart, or scrape past each other. These movements and interactions of tectonic plates are responsible for the formation of many landforms, such as mountains, and for most earthquakes and volcanic eruptions.

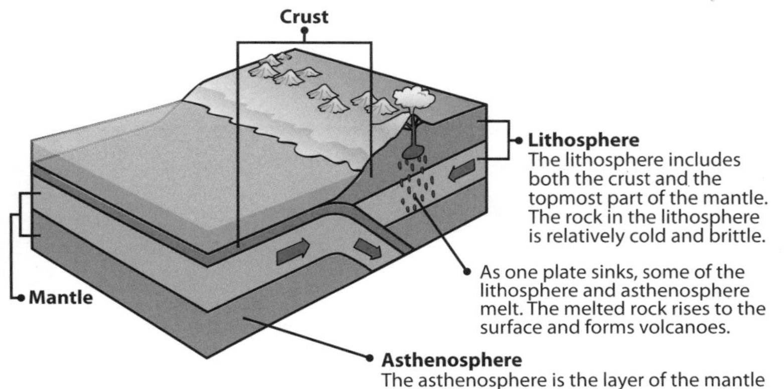

2. Based on the diagram, where does melting of rock happen?

   (1) on Earth's surface
   (2) within the crust
   (3) in the mantle
   (4) at the surface of an oceanic plate
   (5) in the deepest parts of the core

3. Based on the information, what can you conclude about volcanoes?

   (1) Volcanoes are common where two plates collide.
   (2) Volcanoes form when the crust melts.
   (3) Most volcanoes are far from the ocean.
   (4) The rock in a volcano comes mainly from the lowest part of the mantle.
   (5) The tallest volcanoes are found beneath the oceans.

Questions 4 and 5 refer to the following paragraph and diagram.

Water is a powerful force in shaping Earth's surface. Flowing water can pick up and carry rock and soil from one place to another. This process is called erosion. Fast-moving water causes more erosion than slow-moving water, because fast-moving water can carry more rock and soil than can slow-moving water.

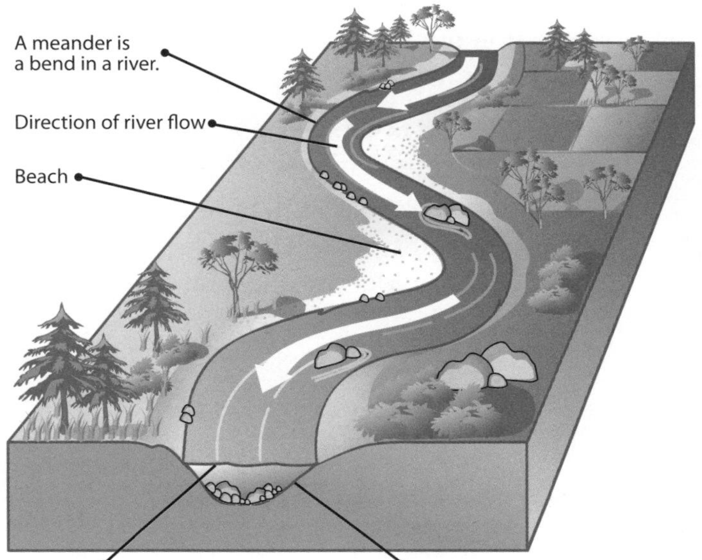

4. Based on the information, at which point does the most erosion probably occur?

   The most erosion occurs

   (1) on the outside of a meander
   (2) in the widest parts of the river
   (3) on the inside of a meander
   (4) in the shallowest parts of the river
   (5) at the river's surface

5. Based on the information, what can you infer about where rock and soil are laid down?

   Rock and soil are laid down where the water

   (1) flows slowest
   (2) is cleanest
   (3) speeds up
   (4) is turbulent
   (5) flows in a straight line

UNIT 2

LESSON 3

Understand Maps and Map Symbols

1 Learn the Skill

Maps provide information about a particular region in a graphic form. No one map can provide all possible information about a particular region. Instead, different maps are used to show specific kinds of information. A map may show roads, bodies of water, landforms, climate patterns, types of vegetation or rock, place names, or political borders. Learning how to **understand maps and map symbols** will allow you to access and interpret a wide variety of information.

2 Practice the Skill

By mastering the skill of understanding maps and map symbols, you will improve your study and test-taking skills, especially as they relate to the GED Science Test. Examine the map and strategies below. Then answer the question that follows.

A The focus of this particular map is bodies of water. The rivers might look like roads. Check the map key to understand what the lines and symbols on the map represent.

B The compass rose tells how the map is positioned relative to the four main compass directions. The scale bar helps you determine the size of landforms or the distance between places shown on the map. The scale bar tells the relationship between distances on the map and actual distances.

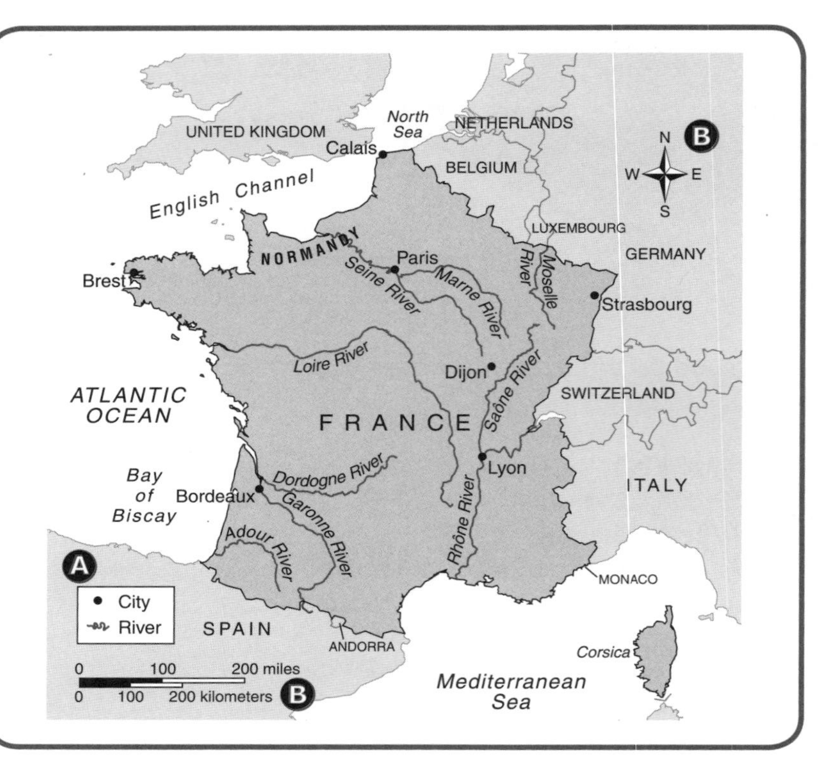

MAKING ASSUMPTIONS

Depending on the purpose of a map, lines that look like boundaries could represent other features, such as areas with the same elevation or the boundaries of different biomes.

1. Which of the following pieces of information could you get from this map?

   From this map you could determine the

   (1) distance between Paris and Berlin
   (2) major east-to-west highway routes in France
   (3) biomes of France
   (4) longest river in France
   (5) population of Paris

③ Apply the Skill

Directions: Choose the one best answer to each question.

Questions 2 through 5 refer to the following paragraph and map.

The map below shows the major biomes found in the United States. A biome is a large region of Earth with a certain type of climate and vegetation, or plant life. Biomes are generally named for their most common type of vegetation. For example, grasses are the most common plants in a grassland biome.

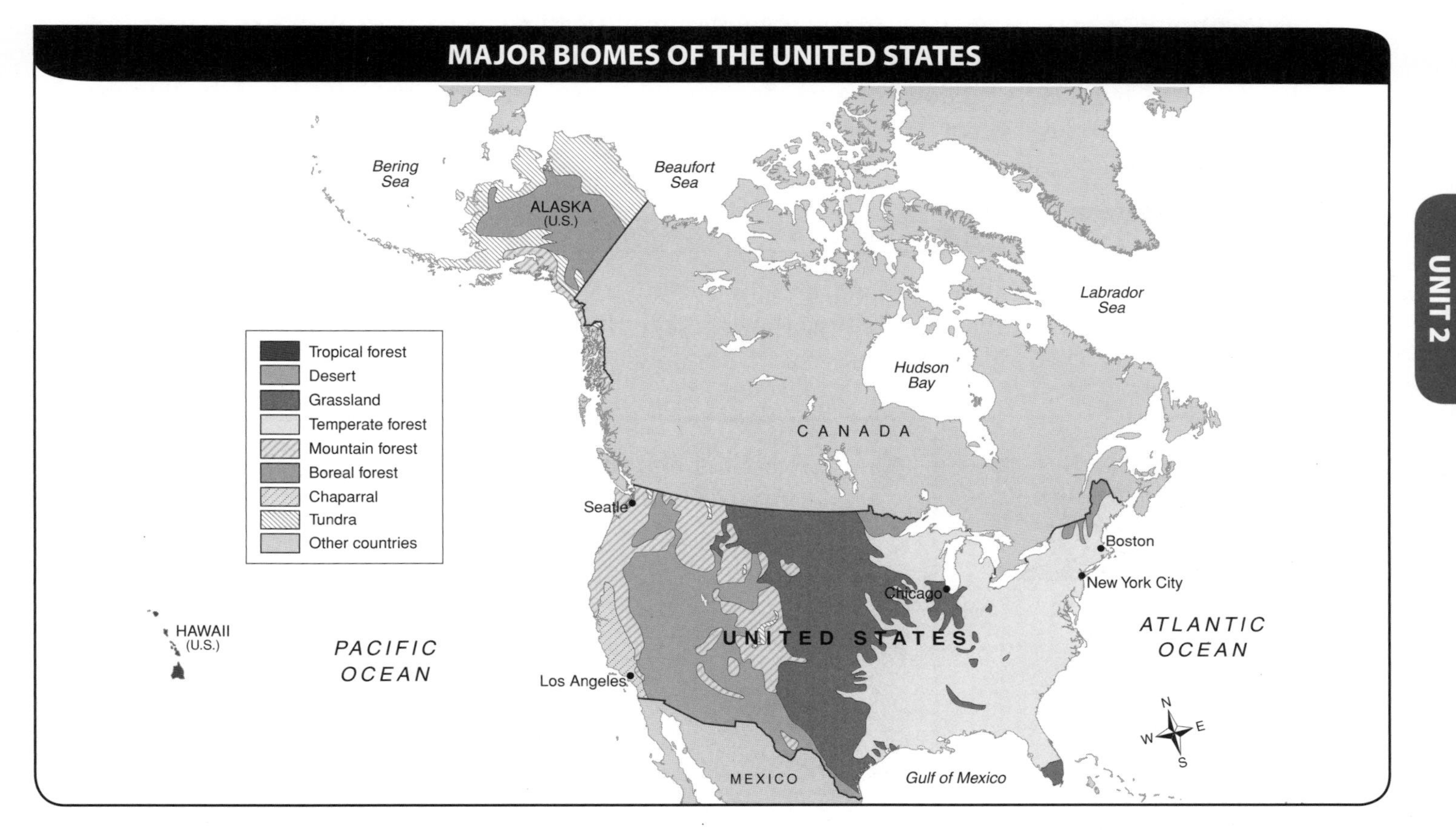

2. In which biome is New York City located?

   (1) desert
   (2) grassland
   (3) temperate forest
   (4) boreal forest
   (5) tundra

3. Which biome is found directly north of the temperate forest biome in some areas?

   (1) desert
   (2) tropical forest
   (3) chaparral
   (4) tundra
   (5) boreal forest

4. The information in the map supports which of the following ideas?

   (1) State borders follow biome borders.
   (2) Seattle and Los Angeles have different climates.
   (3) New York City gets more rain than Boston.
   (4) Larger states contain more biomes.
   (5) The United States is located in the temperate forest biome.

5. Based on the map, on average which two biomes are most likely the coldest?

   (1) desert and grassland
   (2) grassland and temperate forest
   (3) tundra and boreal forest
   (4) mountain forest and tropical forest
   (5) mountain forest and grassland

Interpret Physical and Topographic Maps

1 Learn the Skill

Maps that show natural features such as mountains, lakes, rivers, and oceans are **physical maps**. **Topographic maps** display these as well as human-made features, such as roads and boundary lines. To **interpret physical and topographic maps**, you must understand the use of shading and contour lines to show landforms in different ways. For example, a physical map uses colors and shading to show the general elevation of landforms. A topographic map uses contour lines, which are lines that connect points on a map with the same elevation.

2 Practice the Skill

By mastering the skill of interpreting physical and topographic maps, you will improve your study and test-taking skills, especially as they relate to the GED Science Test. Examine the maps and strategies below. Then answer the question that follows.

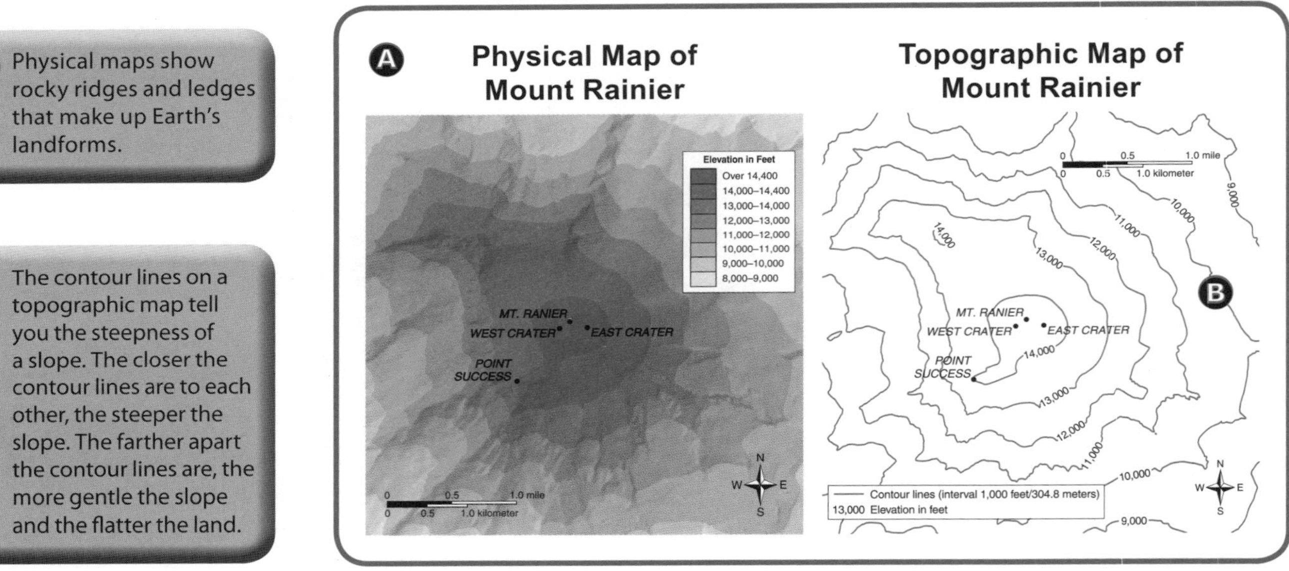

A Physical maps show rocky ridges and ledges that make up Earth's landforms.

B The contour lines on a topographic map tell you the steepness of a slope. The closer the contour lines are to each other, the steeper the slope. The farther apart the contour lines are, the more gentle the slope and the flatter the land.

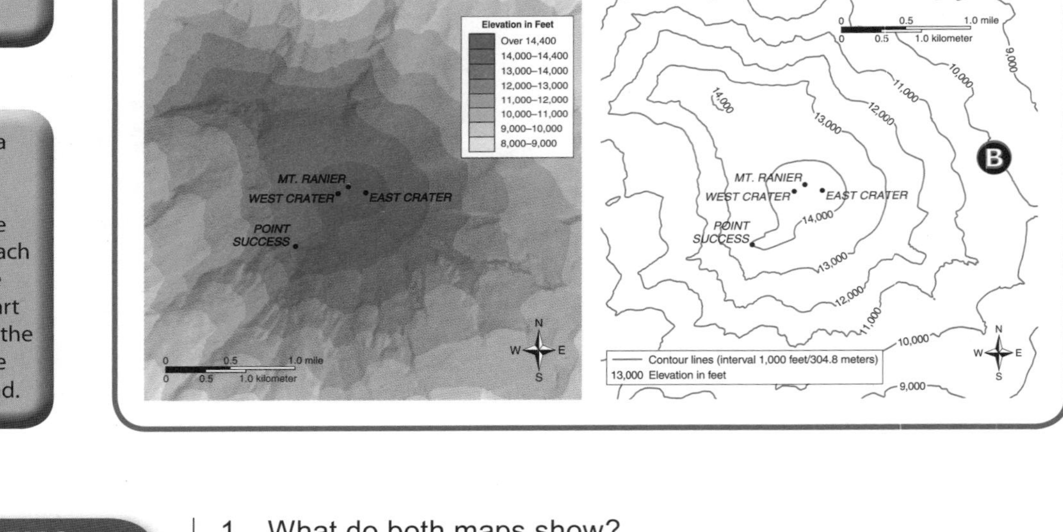

TEST-TAKING TIPS

The most important factor in reading a map is understanding its key. When reading a topographic map, look in the key for the "contour interval." It tells you the elevation between two contour lines.

1. What do both maps show?

   (1) boundaries
   (2) contour lines
   (3) elevation
   (4) human-made features
   (5) forests

3 Apply the Skill

Directions: Choose the one best answer to each question.

Questions 2 through 6 refer to the following paragraph and maps.

The maps below show Mount Saint Helens, which is part of the Cascade Mountains in Washington State. On May 18, 1980, a volcanic eruption occurred there that caused the peak of the mountain to collapse.

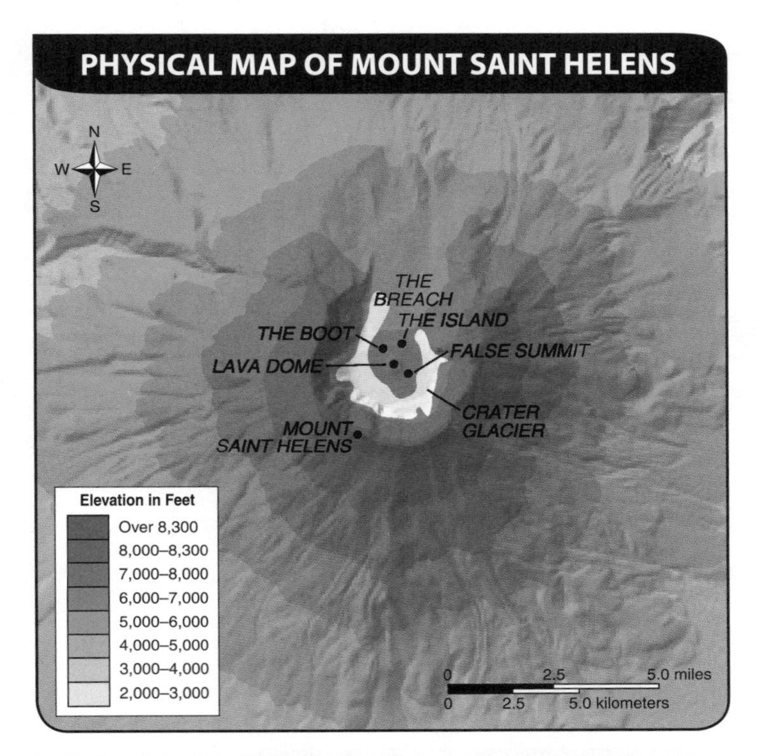

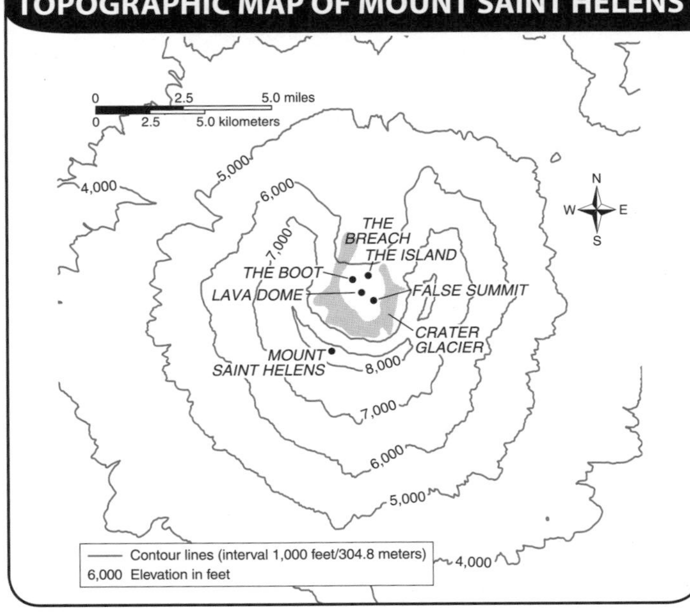

2. Based on the information in the paragraph and in the maps, which best describes the appearance of Mount Saint Helens today?

   (1) a glacier
   (2) a crater
   (3) a peak
   (4) a fault
   (5) a lake

3. At approximately what elevation are the boot, lava dome, and false summit?

   (1) 8,000 feet
   (2) 7,500 feet
   (3) 7,000 feet
   (4) 6,500 feet
   (5) 5,500 feet

4. Which point on the maps has the highest elevation?

   (1) Crater Glacier
   (2) Lava Dome
   (3) False Summit
   (4) Mount Saint Helens
   (5) The Breach

5. If you walked across this landform from south to north, what would you notice about the elevation?

   You would travel

   (1) down a steep hill, and then up a shallow hill
   (2) along level ground, and then downhill
   (3) up and down several large hills
   (4) up about 4,000 feet in elevation, and then down about 4,000 feet
   (5) downhill the whole way

6. Based on the maps, what is the difference in elevation between Mount Saint Helens and the island?

   (1) about 500 feet
   (2) about 1,000 feet
   (3) about 2,000 feet
   (4) about 6,000 feet
   (5) about 8,000 feet

UNIT 2

LESSON 5

Interpret Flowcharts

1 Learn the Skill

Flowcharts are diagrams that show the steps in a process. Many flowcharts use boxes and arrows to show the relationship between each step. When you **interpret flowcharts**, you study the steps in a process, such as how a recycling program works or how to fix a problem with your computer.

2 Practice the Skill

By mastering the skill of interpreting flowcharts, you will improve your study and test-taking skills, especially as they relate to the GED Science Test. Examine the flowchart and strategies below. Then answer the question that follows.

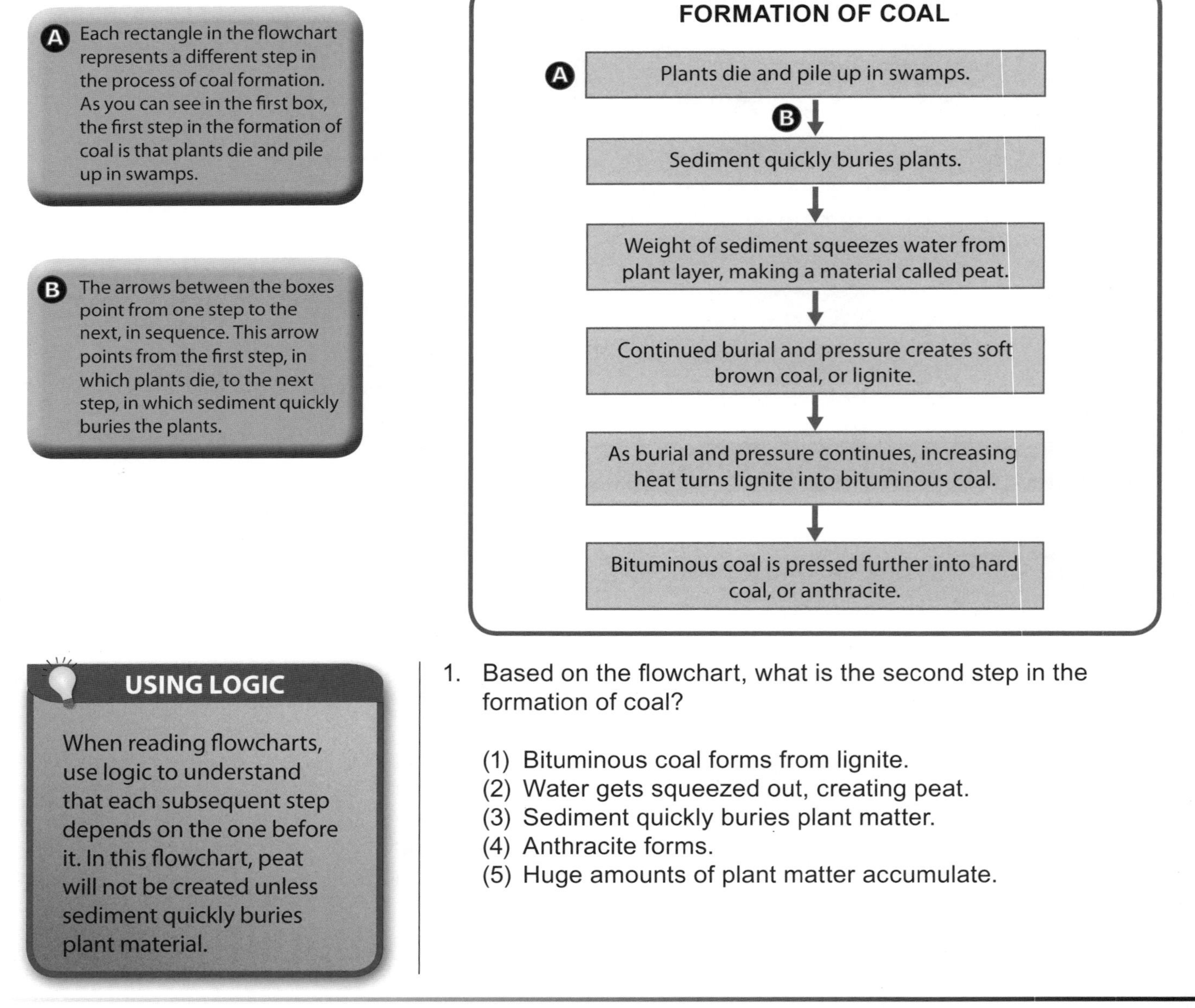

USING LOGIC

When reading flowcharts, use logic to understand that each subsequent step depends on the one before it. In this flowchart, peat will not be created unless sediment quickly buries plant material.

1. Based on the flowchart, what is the second step in the formation of coal?

   (1) Bituminous coal forms from lignite.
   (2) Water gets squeezed out, creating peat.
   (3) Sediment quickly buries plant matter.
   (4) Anthracite forms.
   (5) Huge amounts of plant matter accumulate.

3 Apply the Skill

Directions: Choose the one best answer to each question.

Questions 2 through 6 refer to the following paragraph and flowchart.

The flowchart below shows the rock cycle. A cycle is a series of steps that repeats again and again. In the rock cycle, the conditions surrounding rocks, such as exposure to weather or location beneath Earth's surface, can change rock types. For example, if an igneous rock is weathered down into sediment, it may eventually become a sedimentary rock. Similarly, if an igneous rock is exposed to high heat and pressure beneath Earth's surface, it may change into a metamorphic rock.

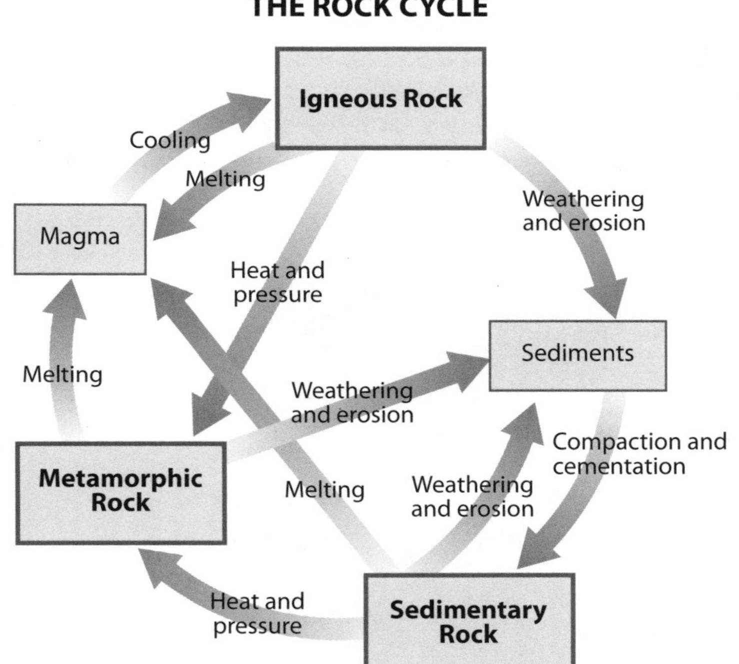

2. Based on the paragraph and flowchart, what turns igneous rock into metamorphic rock?

   (1) compaction and cementation
   (2) weathering and erosion
   (3) heat and pressure
   (4) cooling
   (5) melting

3. The paragraph and flowchart describe a process that is a cycle. What does this tell you about the process of rock formation?

   (1) The process has a beginning and an end.
   (2) It is an ongoing process.
   (3) The process happened millions of years ago.
   (4) It is a process that flows in one direction.
   (5) The process adds rock to the environment.

4. Based on the flowchart, which of the following statements is true?

   (1) Sedimentary rock can never become metamorphic rock.
   (2) Rock never gets hot enough to melt.
   (3) Igneous rock always becomes metamorphic rock.
   (4) Each type of rock can become any other type of rock.
   (5) Igneous rock never becomes metamorphic rock.

5. Based on the flowchart, which of the following processes results in the formation of magma?

   (1) heat and pressure on metamorphic rock
   (2) melting of various forms of rock
   (3) heat and pressure on sedimentary rock
   (4) weathering of igneous rock
   (5) compaction and cementation of sediments

6. Which of the following statements best describes why metamorphic rocks are most commonly found in mountain ranges?

   (1) Mountain ranges contain large amounts of magma.
   (2) Erosion occurs more rapidly on tall mountains.
   (3) Sediment builds up quickly at the base of a mountain range.
   (4) Melting occurs deep within most mountain ranges.
   (5) The processes that produce mountains generate considerable heat and pressure.

UNIT 2

LESSON 6

Compare and Contrast Visuals

1 Learn the Skill

When you compare and contrast, you look for similarities and differences among objects, processes, events, and ideas. When you **compare and contrast visuals**, you identify the similarities and differences in tables, charts, graphs, diagrams, and illustrations.

2 Practice the Skill

By mastering the skill of comparing and contrasting visuals, you will improve your study and test-taking skills, especially as they relate to the GED Science Test. Examine the information, diagrams, and strategies below. Then answer the question that follows.

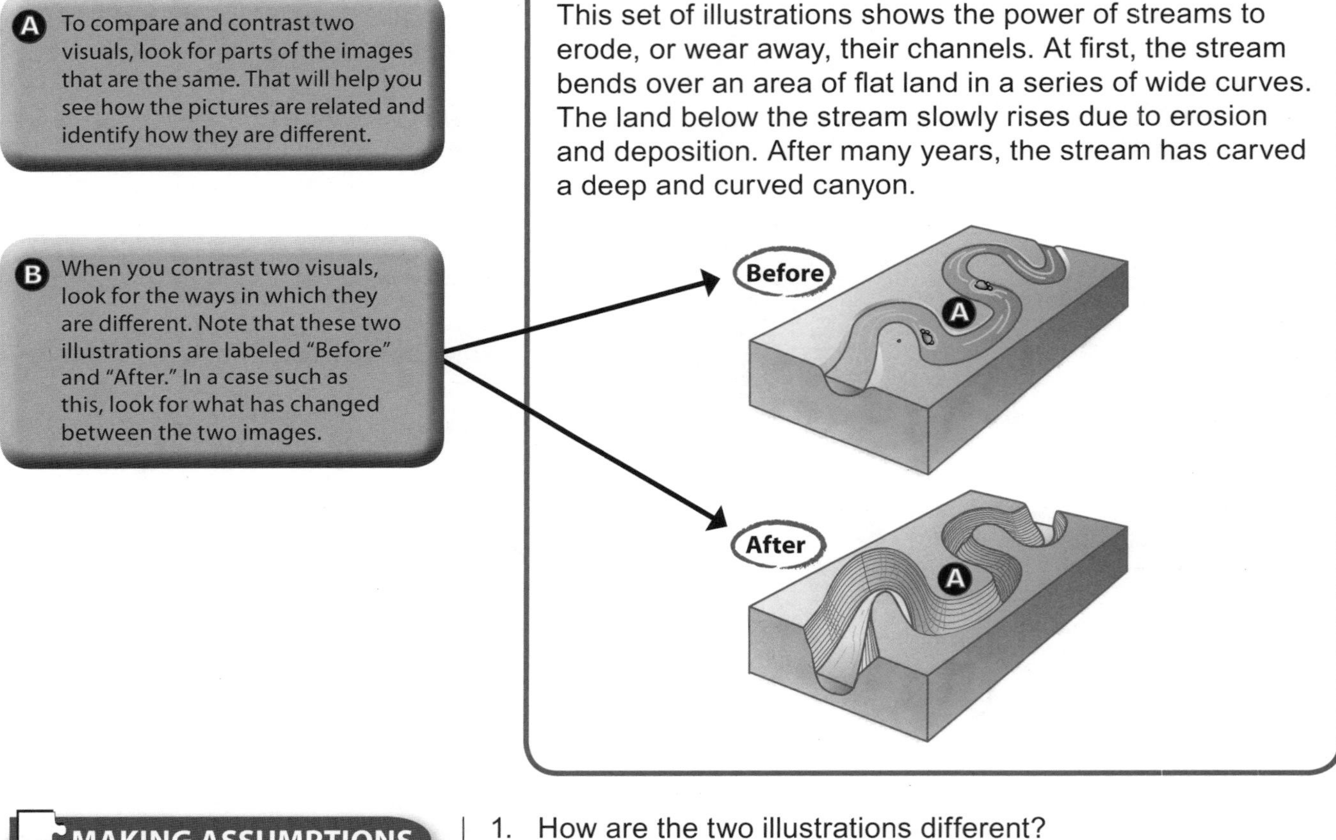

A To compare and contrast two visuals, look for parts of the images that are the same. That will help you see how the pictures are related and identify how they are different.

B When you contrast two visuals, look for the ways in which they are different. Note that these two illustrations are labeled "Before" and "After." In a case such as this, look for what has changed between the two images.

This set of illustrations shows the power of streams to erode, or wear away, their channels. At first, the stream bends over an area of flat land in a series of wide curves. The land below the stream slowly rises due to erosion and deposition. After many years, the stream has carved a deep and curved canyon.

MAKING ASSUMPTIONS

You might assume that a river doesn't change much over time, but it does. If the land over which a river flows changes in some way, then the river will flow differently.

1. How are the two illustrations different?

   In the second illustration, the stream

   (1) flows in the opposite direction
   (2) has disappeared
   (3) is surrounded by flatter land
   (4) has cut a deep canyon
   (5) follows a different course

③ Apply the Skill

Directions: Choose the one best answer to each question.

Questions 2 and 3 refer to the following paragraph and maps.

Longshore currents run parallel to the coast. These currents carry sediment as they move. Wherever they deposit sediment, they can create new landforms.

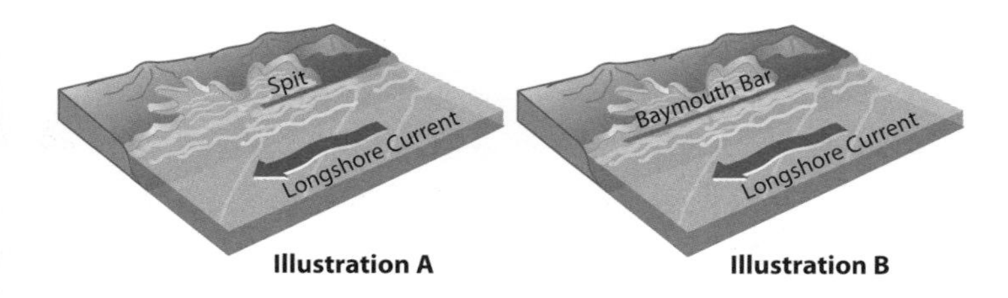

2. Illustrations A and B show the same area at two different times. Illustration A occurred before Illustration B. What did the area most likely look like before Illustration A?

   (1) The spit was larger.
   (2) The spit did not exist.
   (3) The longshore current ran in the opposite direction.
   (4) The baymouth bar was larger.
   (5) The water was deeper.

3. Compare what occurred in Illustration B to what occurred in Illustration A.

   The current has

   (1) continued to build the spit until it has become a baymouth bar
   (2) washed away the spit and built a baymouth bar in its place
   (3) reversed direction and built a baymouth bar
   (4) built a spit where a baymouth bar used to be
   (5) made the offshore water deeper, destroying the spit

Questions 4 and 5 refer to the following paragraph and graphs.

When it rains, some water falls into bodies of water. Water that falls on land becomes runoff in a watershed. A watershed includes all the water that runs off the land in streams and eventually flows into a river, lake, or ocean. The speed of runoff depends on many things, including the slope of the land, the size of the channels, and the landforms and features of the watershed. Soil quality also affects runoff. Permeable surfaces, such as healthy soil and wetlands, can absorb water, while impermeable surfaces, such as asphalt and compacted soil, create more runoff. The two graphs show different streams' water volumes after a storm.

RUNOFF IN TWO STREAMS AFTER A STORM

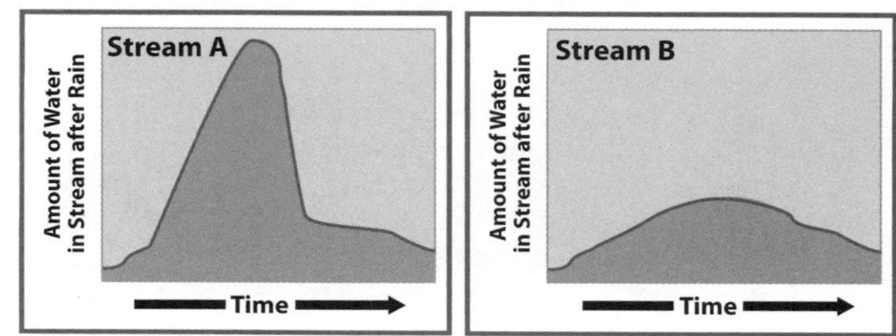

4. Which of the following can you conclude based on the information?

   Stream A

   (1) is part of a different watershed than Stream B
   (2) absorbs rainwater more easily than Stream B
   (3) discharges into a different river than Stream B
   (4) is part of a watershed with less absorbent soil than Stream B
   (5) is wider than Stream B

5. In which way do streams A and B differ?

   (1) More rainwater fell directly into Stream B.
   (2) Stream B is probably next to a parking lot.
   (3) Stream A is probably beside a wetland.
   (4) Stream B has an overall larger volume of water.
   (5) Stream A accumulated more runoff during the storm.

LESSON 7

Determine Fact and Opinion

1 Learn the Skill

When you read text or the captions on photographs, illustrations, diagrams, or cartoons, you need to know how to evaluate them. That's why it is important to know how to **determine fact and opinion**. A **fact** is something that can be proved true or false. An **opinion** is a point of view that cannot be proved or supported with facts. Sometimes, text that is opinion appears to be fact. Determining the difference can be useful.

2 Practice the Skill

By mastering the skill of determining fact and opinion, you will improve your study and test-taking skills, especially as they relate to the GED Science Test. Read the excerpt and strategies below. Then answer the question that follows.

A Never trust any writing based solely on its quality. Terms such as *postulate* and *pervade* may sound impressive, but they do not prove that the text is entirely factual. Always check the source.

"The Big Bang Model ... postulates that 12 to 14 billion years ago, the portion of the universe we can see today was only a few millimeters across. It has since expanded from this hot dense state into the vast and much cooler cosmos we currently inhabit. We can see remnants of this hot dense matter as the now very cold cosmic microwave background radiation which still pervades the universe and is visible to microwave detectors as a uniform glow across the entire sky."

From *The National Aeronautics and Space Administration (NASA.gov)*

B To answer the question, first ask yourself, "What would make readers trust this writing?" Then look at the source of the writing. In this case, you can find the source in the credit line. The credit line signifies that the material comes from the National Aeronautics and Space Administration, which is an official, trusted source of information.

MAKING ASSUMPTIONS

Never assume that a report is unbiased and only contains facts just because it comes from an official source. Look closely at the source to determine whether there might be a political motive for the report or whether the organization behind the report is promoting a certain point of view.

1. Why is this excerpt easy to accept as fact?

   (1) The writing is intelligent.
   (2) The passage is from a respected organization that uses valid research.
   (3) Everyone agrees with the Big-Bang Model.
   (4) All models are considered fact.
   (5) The passage contains many specific details.

③ Apply the Skill

Directions: Choose the one best answer to each question.

Questions 2 and 3 refer to the following paragraph.

> Many scientists believe that our solar system was formed by a solar nebula. A solar nebula is a large cloud of rotating gases and dust from which the solar system formed about 4.6 billion years ago. Research shows that the gases and dust cooled over time, causing the cloud to condense and collect. Eventually, the cloud collapsed and flattened into a disk. Matter inside the disk collected to form the planets and the sun.
>
> From *The National Aeronautics and Space Administration (NASA.gov)*

2. Which of the following word or words indicates that the passage is mostly fact?

   (1) "scientists believe"
   (2) "4.6 billion years"
   (3) "over time"
   (4) "research shows"
   (5) "eventually"

3. Which of the following is an opinion based on this paragraph?

   (1) The solar nebula created our solar system.
   (2) The solar nebula theory is the best explanation for how the solar system was formed.
   (3) Our solar system formed more than four billion years ago.
   (4) The sun formed from a collection of matter inside the solar nebula.
   (5) Our solar system is a disk-like shape.

Questions 4 and 5 refer to the following timeline.

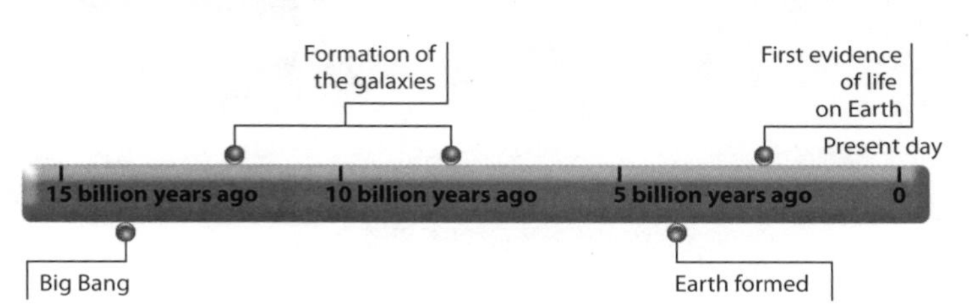

4. Which statement below is a fact based on the timeline?

   (1) The Big Bang is an unimportant event in the history of the universe.
   (2) Galaxies began to form approximately 1 billion years after the Big Bang.
   (3) The next step on the timeline will be a second Big Bang.
   (4) More events should be added to the timeline.
   (5) The Earth formed before galaxies did.

5. Which of the following information would be most reliable to use in constructing a timeline like the one above?

   (1) dates taken from journals written by early scientists
   (2) data collected from science-fiction movies
   (3) observations from astronauts who have explored space
   (4) data based on research performed by respected astronomers
   (5) information collected from interviewing astronomy students

UNIT 2

LESSON 8

Evaluate Information

UNIT 2

1 Learn the Skill

When you **evaluate information**, you judge it in several ways. You assess its effectiveness in communicating a main idea, data, or a point of view. You also assess the source as well as the quality and accuracy of the data incorporated. Evaluating information can help you get the most from a variety of sources.

2 Practice the Skill

By mastering the skill of evaluating information, you will improve your study and test-taking skills, especially as they relate to the GED Science Test. Read the excerpt and strategies below. Then answer the question that follows.

A Note that there is little opinion about Pluto's demotion in the article. The author neither agrees nor disagrees with the decision to call Pluto a dwarf planet. However, there is strong opinion about the emotion involved in the debate. Pay attention to any opinion that seems to slip into an article or essay. Too much opinion can indicate that a source is unreliable.

B Information is only as good as its source. Certain newspapers and magazines, such as the *New York Times*, *Scientific American*, and *Science*, have a strong, well-earned reputation for fair and accurate reporting.

PLUTO IS DEMOTED TO 'DWARF PLANET'

"After years of wrangling and a week of bitter debate, astronomers voted on a sweeping reclassification of the solar system. In what many of them described as a triumph of science over sentiment, Pluto was demoted to the status of 'dwarf planet.' . . .

"It has long been clear that Pluto . . . stood apart from the previously discovered planets. Not only was it much smaller . . . but its elongated orbit tilted with respect to the other planets and it goes inside the orbit of Neptune part of its 248-year journey around the Sun. . . .

"Two years ago, the International Astronomical Union appointed a working group of astronomers to come up with a definition that would resolve the tension. . . ."

From the *New York Times*, August 24, 2006

TEST-TAKING TIPS

If you are trying to evaluate sources of information, it helps to be familiar with newspapers, magazines, and Web sites known for being good and reliable sources of news and information. You also should be familiar with those sources that have a political slant or bias.

1. Which characteristic of this article helps you evaluate it as a valid source of information?

   (1) It reports on a recent event.
   (2) It mentions the International Astronomical Union (IAU).
   (3) It is about a subject often discussed in the news.
   (4) It is printed in a respected newspaper.
   (5) It is written with correct spelling and grammar.

③ Apply the Skill

Directions: Choose the one best answer to each question.

Questions 2 and 3 refer to the following excerpt.

"The International Astronomical Union (IAU) Resolution means that the Solar System consists of eight planets: Mercury, Venus, Earth, Mars, Jupiter, Saturn, Uranus, and Neptune. A new distinct class of objects called dwarf planets was also decided on. It was agreed that planets and dwarf planets are two distinct classes of objects. The first members of the dwarf planet category are Ceres, Pluto, and Eris. . . ."

From the International Astronomical Union's Web site

2. Why would the International Astronomical Union be a good source for information about the reclassification of Pluto?

   (1) Astronomers are experts when it comes to the solar system.
   (2) The IAU agreed with the decision.
   (3) As an international organization, it would not have local bias.
   (4) The IAU made the decision.
   (5) The IAU is probably the only organization with updated information about Pluto.

3. There were many stories about the change in the status of Pluto when it occurred. Which of the following would be most reliable source for a story about Pluto?

   (1) *Science Fiction Weekly*
   (2) the Hayden Planetarium
   (3) *Travel and Leisure* magazine
   (4) a newsletter published by the Museum of the American Indian
   (5) the Web site of the U.S. Environmental Protection Agency

Questions 4 and 5 refer to the following excerpts.

A. From *Harold Kinder, Ph.D. candidate, posted at superplanetguy.blogspot.com, December 2006*

"Oh, I just heard that the IAU demoted Pluto. Good for them! It should never have been a planet in the first place. They made the right decision as far as I'm concerned. It's a rock out in space—basically a wayward asteroid. I could pluck out about 20 such rocks in the belt and call them planets too."

B. From *The Encyclopedia of Science,* 1988

"Pluto is the ninth planet in the solar system, and the only rocky planet of the outer planets. This makes it different from the gas giants beyond the asteroid belt. It has an orbit of 248 years, and its orbit swoops out of alignment with the other planets. Pluto's moon, Charon, was discovered in 1978."

4. Suppose you are writing a report about the characteristics of Pluto. Which of the following statements best explains why Excerpt A would not be an appropriate source?

   (1) It is not written by a professor.
   (2) It gives mainly opinions.
   (3) It was published online.
   (4) It is very short.
   (5) There is no information about the writer's qualifications.

5. Which of the following statements best explains why Excerpt B would not be an appropriate source for the report?

   (1) It is not based on facts.
   (2) It gives information about Pluto's moons.
   (3) It compares Pluto to gas giants.
   (4) It was written more than 20 years ago.
   (5) It does not give information about the scientist who discovered Pluto.

LESSON 9

Identify Problem and Solution

1 Learn the Skill

Sometimes, an author may organize a subject around a problem and the way in which it was solved. When you **identify problem and solution**, you locate the problems presented in the text and evaluate the possible solutions that are outlined by the author. This will help you understand more about the subject and its history.

2 Practice the Skill

By mastering the skill of identifying problem and solution, you will improve your study and test-taking skills, especially as they relate to the GED Science Test. Read the paragraph and strategies below. Then answer the questions that follow.

A One problem is clearly identified in both the article's title and the paragraph's first two sentences. However, there are other problems mentioned in the paragraph. You must decide which is the main problem.

B There is a solution presented in this paragraph, but it doesn't solve all the problems. The paragraph explains how it solves problems in Asia, but not in Africa.

LACK OF CLEAN WATER HARMS MANY

Clean water is essential for every person. But more than 1 billion people worldwide have no access to clean water, according to the United Nations children's agency, UNICEF. Each year, 1.5 million children under age 5 die from disease due to unsanitary conditions and lack of access to safe drinking water. The problem is especially serious in certain parts of Asia and Africa. However, the UNICEF report calls South Asia a success story, with improvements in sanitation and increased access to clean water over the last couple of decades. Projects in sub-Saharan Africa have met with less success. In some areas, especially those with armed conflicts, only a small percentage of people have access to clean water.

TEST-TAKING TIPS

Try drawing a flowchart to organize problems and solutions. If you identify a problem, write it down. Then read on to see if there is a solution to the specific problem. Sometimes, a text will mention several solutions; sometimes, though, no solution may be mentioned. Solutions should be considered critically so as to avoid causing larger problems or side effects.

1. Which is the main problem, as stated in the paragraph?

   (1) armed conflicts in Africa
   (2) lack of sanitation and clean water
   (3) interference from UNICEF
   (4) an increase in population
   (5) unsuccessful projects in the Sahara

2. What is a possible solution in the paragraph for the situation in sub-Saharan Africa?

   (1) a groundbreaking report by UNICEF
   (2) vaccination against disease
   (3) moving to South Asia
   (4) improvements in sanitation
   (5) providing clean water to armies

③ Apply the Skill

Directions: Choose the one best answer to each question.

Questions 3 and 4 refer to the following paragraph and diagram.

Earth's supply of fresh water is an important natural resource. There are freshwater shortages in many parts of the world. As the world's population grows, even greater supplies will be needed. Therefore, it is important that we protect our clean water resources. Cleaning up polluted lakes and rivers is one aspect of protecting freshwater supplies.

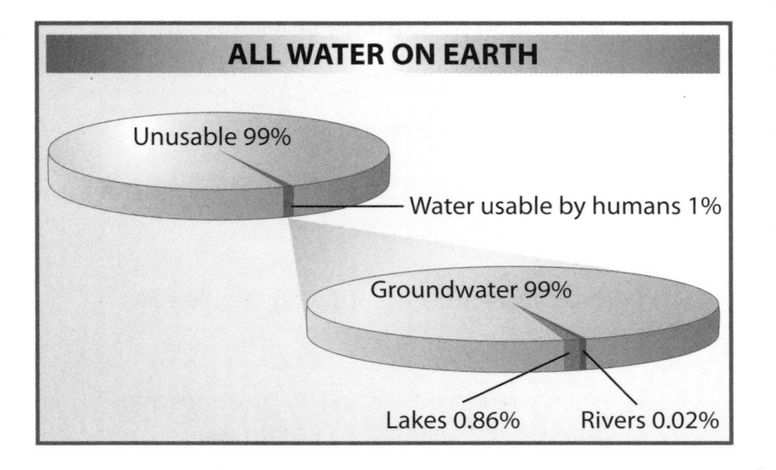

3. The diagram shows the distribution of Earth's water. All living things on Earth need fresh water. Interpret the diagram to identify the problem with Earth's water supply.

   Earth's water supply

   (1) is too large
   (2) does not contain enough salt water
   (3) has most of its fresh water in swamps
   (4) has more fresh groundwater than salty ocean water
   (5) has a relatively small proportion of fresh water

4. Based on the paragraph and the diagram, which of these is a solution to the problem?

   (1) finding new supplies of fresh water
   (2) turning frozen ice caps into liquid fresh water
   (3) protecting scarce supplies of liquid fresh water
   (4) drinking salt water from the ocean instead of fresh water
   (5) drinking fresh water from swamps

Questions 5 and 6 refer to the following paragraph.

Most of Earth's water supply is salt water, which people cannot drink. What if we could turn salt water into fresh water? Actually, we can. This process is called desalination, and it provides water to people in some parts of the world where fresh water is scarce. There are various methods of desalination. In one, salt water is heated until the fresh water evaporates, leaving the salt behind. The water vapor then cools and condenses. The fresh water is collected in another chamber. In the United States, a few towns in California and Florida are already using desalination to meet their freshwater needs.

5. Based on the information, what is the main problem?

   (1) lack of fresh water in some places
   (2) towns in California and Florida without salt water
   (3) salt water overheating and causing fresh water
   (4) too much of our water supply is salt water
   (5) the relative difficulty of desalination

6. Desalination is given as a potential solution to the problem in the paragraph. Which of the following additional pieces of information would best help you evaluate whether this solution could be widely useful?

   (1) the number of towns currently using desalination
   (2) the history of the invention of desalination
   (3) the costs and efficiency of desalination processes
   (4) the names of people using desalination
   (5) the amount of salt water present on Earth

UNIT 2

Unit 2 Review

The Unit Review is structured to resemble the GED Science Test. Be sure to read each question and all possible answers very carefully before choosing your answer.

To record your answers, fill in the numbered circle that corresponds to the answer you select for each question in the Unit Review.

Do not rest your pencil on the answer area while considering your answer. Make no stray or unnecessary marks. If you change an answer, erase your first mark completely.

Mark only one answer space for each question; multiple answers will be scored as incorrect.

Sample Question

If you wanted to better understand the elevation of mountainous areas in the United States, which source would be best to use?

(1) a bar graph showing temperature and precipitation in different states
(2) a 3-dimensional diagram showing the internal structure of a mountain range
(3) a scientific study on the effects of greenhouse gases around the world
(4) a physical map of the United States
(5) a newspaper editorial about the importance of conservation in mountainous areas

①②③●⑤

Directions: Choose the one best answer to each question.

Questions 1 and 2 refer to the following paragraph and diagrams.

Sinkholes are depressions that can open up suddenly in the ground. They usually form in areas of soft limestone rock. The process begins when slightly acidic rainfall seeps into the ground. It slowly wears away the rock, forming caves. Sinkholes then can form, as shown in the diagrams below.

Diagram A: Pre-sinkhole

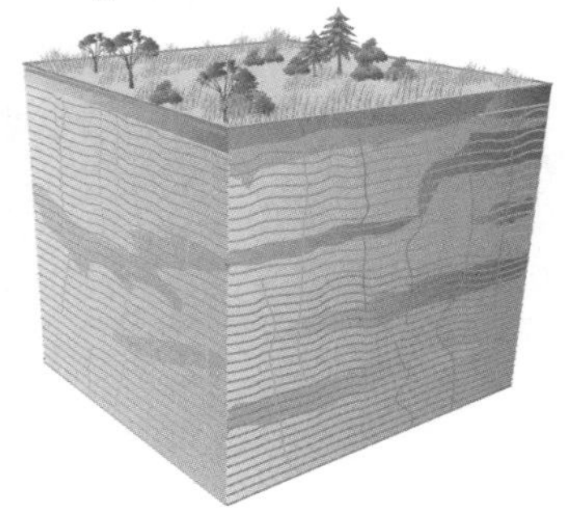

Diagram B: Sinkhole

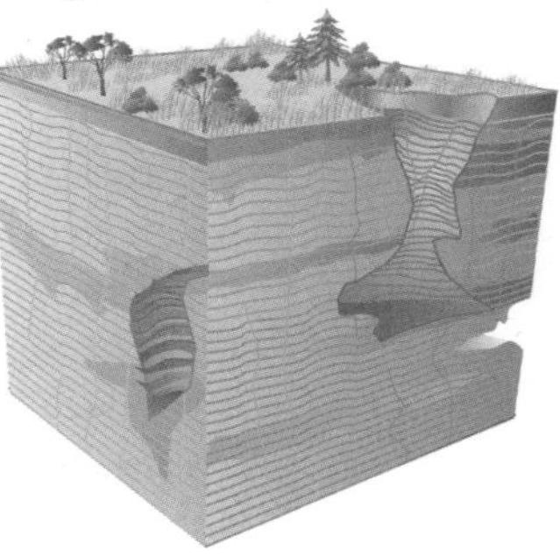

1. What occurred between Diagram A and Diagram B?

(1) Heavy rains collecting on the surface caused the ground to collapse.
(2) The acidic rainfall eating away rock underground caused the collapse.
(3) Soft rock collapsed from the weight of the houses built on it.
(4) Drought caused already dry limestone to become weaker and collapse.
(5) An underground volcano erupted and formed a sinkhole.

①②③④⑤

2. Areas with which of the following characteristics are most likely to experience sinkholes?

(1) hard granite bedrock and dry weather
(2) limestone bedrock and dry weather
(3) slate bedrock and wet weather
(4) limestone bedrock and wet weather
(5) hard granite bedrock and wet weather

①②③④⑤

Questions 3 and 4 refer to the following paragraph.

You are planning to write an article about the International Astronomical Union's (IAU) decision in August 2006 to change the classification of Pluto from a planet to a dwarf planet. You identify the following sources at the library:

1. **Source A (10/2/2006):** a magazine article explaining the decision using quotes from astronomers who are members of the IAU and were involved in the decision.
2. **Source B (9/26/2006):** a newspaper article that neither quotes astronomers nor mentions the IAU. It does include a diagram of the solar system.
3. **Source C (4/18/2004):** an editorial by a famous geochemist about Pluto's atmosphere.

3. Which source(s) would you use in your own article?

(1) source A, because its information comes from astronomers involved in the decision
(2) sources A and C, because they include either thoughts or insights from scientists
(3) source B, because it is from a newspaper rather than a magazine
(4) sources B and C, because they provide scientific facts about Pluto
(5) sources A and B, because they were both published after the decision

①②③④⑤

4. You also find a November 2006 interview with a relative of the scientist who discovered Pluto. She is not an astronomer, but states that she is upset that the IAU changed Pluto's designation. How could this source contribute to your report?

It could

(1) provide unbiased information about the characteristics of Pluto
(2) help you understand why some people disagreed with the IAU's decision
(3) provide details as to why the IAU changed Pluto's designation
(4) help you convince the IAU that its decision was incorrect
(5) show scientific evidence for the existence of Pluto

①②③④⑤

Questions 5 and 6 refers to the following cross-section diagram.

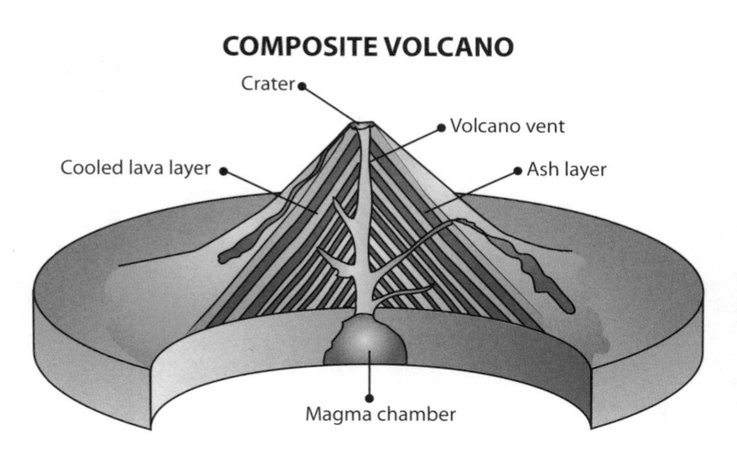

5. Based on the cross-section diagram, which of the following statements is true?

(1) The magma chamber is located on the side of the volcano.
(2) Lava can spill out the sides of the cone during an eruption.
(3) The eruptions from a composite volcano would not be deadly.
(4) Ash and rock are pushed up through the magma chamber in layers.
(5) Composite volcanoes always have one flat, broad side.

①②③④⑤

6. Based on the diagram, what can you infer about composite volcano eruptions?

(1) Composite volcanoes constantly erupt lava.
(2) Composite volcanoes can erupt both lava and ash.
(3) The magma chamber of a composite volcano is not involved in eruptions.
(4) The crater of a composite volcano becomes blocked during an eruption.
(5) A composite volcano erupts in a regular, predictable pattern.

①②③④⑤

UNIT 2

Questions 7 through 9 refer to the following flowchart.

HOW HAIL FORMS

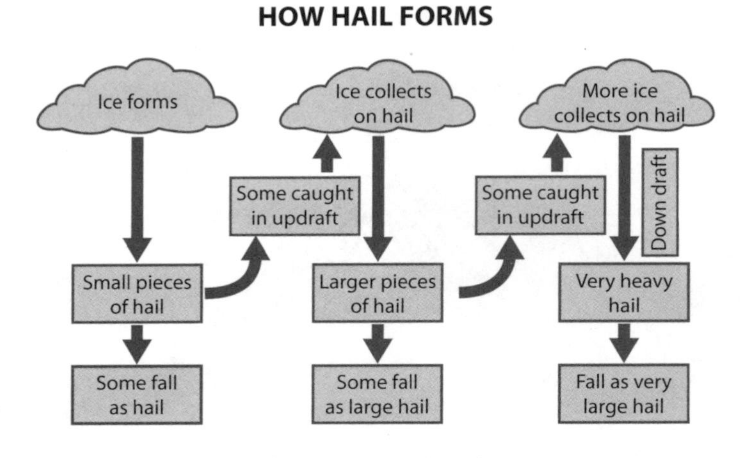

7. Based on the flowchart, what happens to very large hail?

(1) It is caught in an updraft and returns to the clouds.
(2) It falls to Earth in a downdraft.
(3) It turns into clouds.
(4) More ice collects on it.
(5) It damages homes and trees.

①②③④⑤

8. Based on the flowchart, what are you most likely to see during a hailstorm?

(1) snowflakes stuck to large pieces of hail
(2) only very large, very heavy hail
(3) large and small pieces of hail
(4) rain, ice, and snow at the same time
(5) only very small pieces of hail

①②③④⑤

9. Based on the flowchart, which of the following conditions must be present for hail to form?

(1) clouds that are very close to the ground
(2) temperatures high enough to melt ice
(3) strong winds parallel to the ground
(4) clouds that contain both rising and sinking air
(5) strong lightning and thunder

①②③④⑤

Questions 10 and 11 refer to the following paragraph.

Our solar system consists of eight planets and the sun, which is a star. The planets are organized into two categories based on their densities. The first group of planets—Mercury, Venus, Earth, and Mars—is rocky and dense. The second group of planets—Jupiter, Saturn, Uranus, and Neptune—is large and gaseous. The second group of planets also contains ring systems, which are made up of rock and ice particles. An asteroid belt separates the two groups of planets.

10. Which of the following is an opinion?

(1) Because of its similarities to Earth, Mars is likely to contain intelligent life.
(2) The solar system contains eight planets.
(3) Planets closest to the sun are denser than those that are farther away.
(4) All of the planets in the solar system can fall into one of two categories.
(5) Jupiter, Saturn, Uranus, and Neptune all have rings.

①②③④⑤

11. What set of additional information would best support the facts in this passage?

(1) observation from astronauts who have explored space
(2) a chart showing the distance of the planets from the sun
(3) topographic maps of each of the rocky planets
(4) a map of the solar system, showing the locations and orbit patterns of the planets
(5) a list of each planet's moons and their sizes

①②③④⑤

Questions 12 and 13 refer to the following paragraph.

Two scientists moved from the country to a nearby city. They noticed that the temperatures in the city seemed to be much higher than they were in the country. They did some research and learned that the city's temperatures were often higher than the country's temperatures on the same day. After doing more research, they learned that this trend is common in many areas: in general, cities are warmer than the surrounding suburban and rural areas.

12. One scientist hypothesized that the larger number of people living closer together in the city is what causes city temperatures to be higher than temperatures in surrounding areas. How could the scientists best test this hypothesis?

    (1) Measure the temperature at a busy restaurant and a quiet park and compare the data.
    (2) Measure the temperature at a place with many people and then wait until the area is empty to measure it again.
    (3) Measure the temperature of two city parks, one with many people and one with few, at the same time of day.
    (4) Walk around to different places and see how much the temperature changes.
    (5) Visit different city parks and ask people how warm they feel.

①②③④⑤

13. Which of the following is a valid hypothesis, based on the information provided?

    (1) Parks and areas with trees will have lower temperatures than areas without foliage.
    (2) People are causing global warming.
    (3) Cities are warmer because the sun shines more brightly on them.
    (4) Cars and buses cause lower temperatures in certain areas because of how fast they travel.
    (5) Different temperatures prove that certain areas are more populated than others.

①②③④⑤

Questions 14 through 16 refer to the following topographic map.

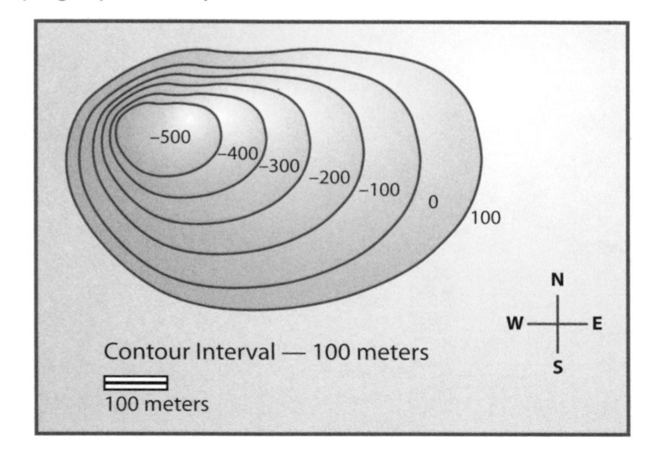

14. What landform does this topographic map most likely show?

    (1) a mountain
    (2) a crater
    (3) a plain
    (4) a highland
    (5) a river

①②③④⑤

15. Based on the map, which of these statements is true of the landform?

    (1) It is more than 1,000 meters in elevation.
    (2) It has a lake in the middle of it.
    (3) It is more than 1,000 meters wide.
    (4) It has at least one steep side.
    (5) It has a river flowing across it.

①②③④⑤

16. A hiker travels across the widest part of the landform from east to west. What will the hiker most likely experience?

    She will travel

    (1) uphill, and then downhill
    (2) downhill, and then uphill
    (3) uphill all the way
    (4) up and down five hills
    (5) up a steep hill at the beginning, and down a shallow hill at the end

①②③④⑤

UNIT 2

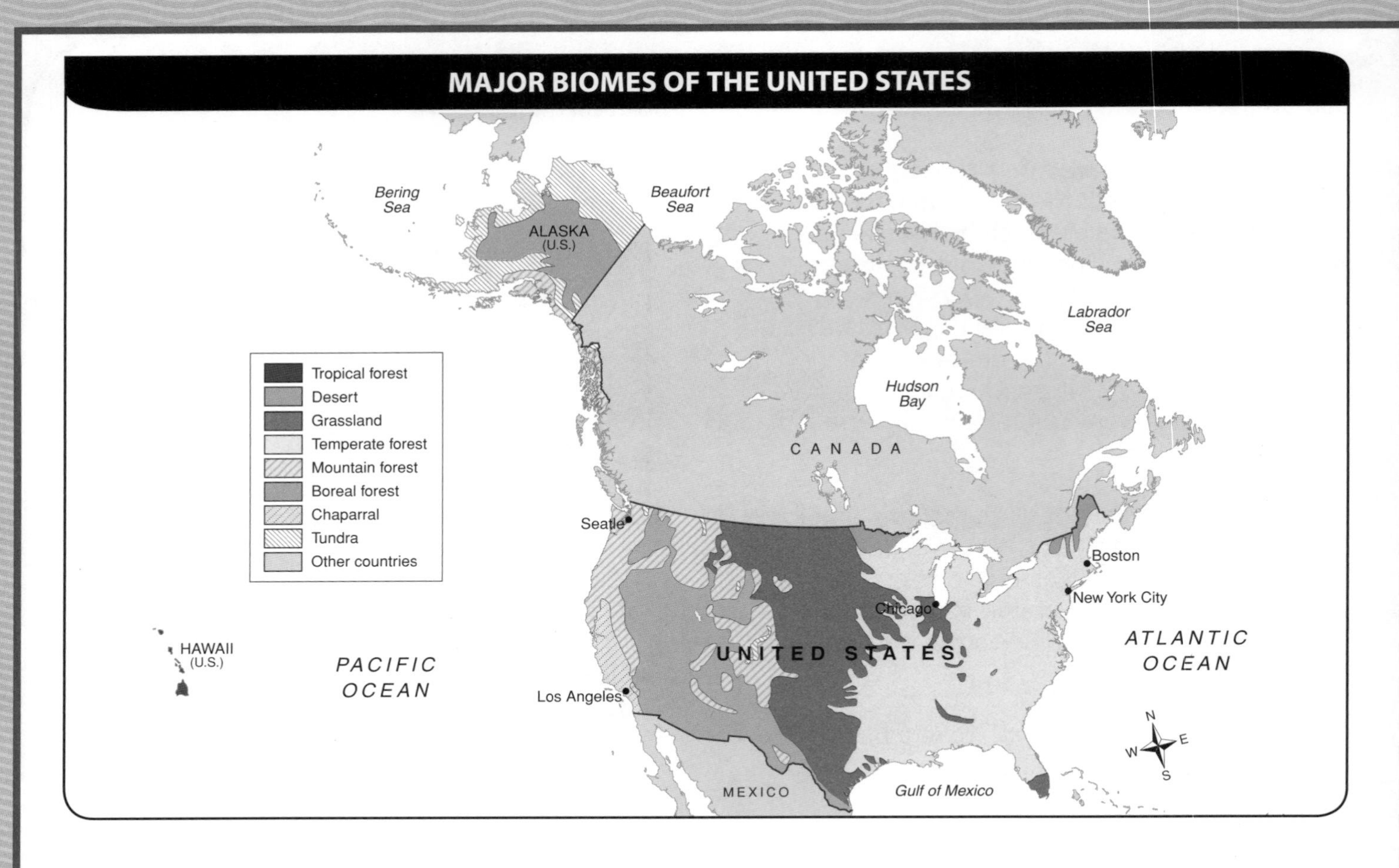

Questions 17 through 20 refer to the above map.

17. In which biome is Chicago located?

    It is in

    (1) the temperate forest biome
    (2) the desert biome
    (3) the grassland biome
    (4) the chaparral biome
    (5) the tropical forest biome

①②③④⑤

18. In which biome is Boston located?

    It is in

    (1) the temperate forest biome
    (2) the tundra biome
    (3) the grassland biome
    (4) the chaparral biome
    (5) the tropical forest biome

①②③④⑤

19. Based on the map, which of the following biomes are most likely warmest?

    (1) chaparral and tundra
    (2) chaparral and desert
    (3) grassland and mountain forest
    (4) tundra and mountain forest
    (5) temperate forest and tundra

①②③④⑤

20. Based on the map, which of the following statements is true?

    (1) Grassland is found only in the northern United States.
    (2) Most parts of the eastern United States are in the boreal forest biome.
    (3) The same biomes are found in Hawaii and in Florida.
    (4) Boreal forest is found only in Alaska.
    (5) Most of the desert in the United States is found in the western and southwestern parts of the country.

①②③④⑤

Questions 21 through 24 refer to the following paragraph and graph.

Fresh water (water that is not salty) is an important natural resource. It is also limited in supply. Many scientists and conservationists are concerned that if we do not begin to conserve fresh water, we will run out. The graph below shows how people in the United States use fresh water.

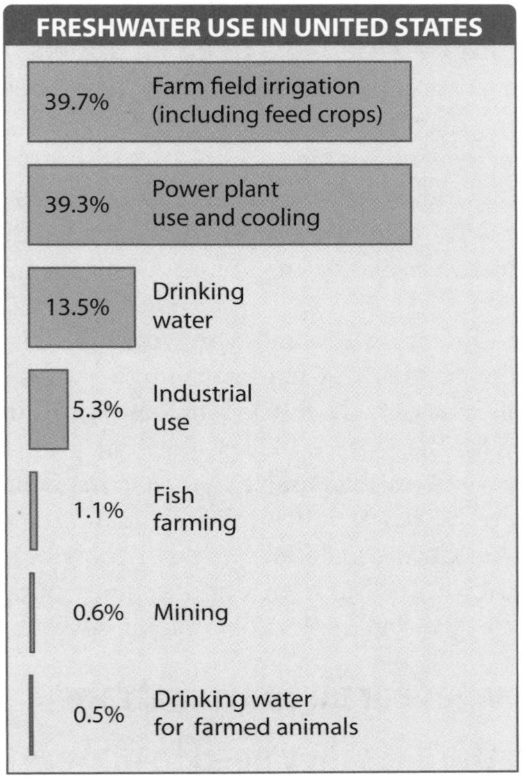

FRESHWATER USE IN UNITED STATES

Percent	Use
39.7%	Farm field irrigation (including feed crops)
39.3%	Power plant use and cooling
13.5%	Drinking water
5.3%	Industrial use
1.1%	Fish farming
0.6%	Mining
0.5%	Drinking water for farmed animals

21. Some people think that the use of fresh, drinkable water to irrigate farm crops is a problem because it wastes drinkable water. Which of the following observations would best support this position?

    (1) Undrinkable water can be used to irrigate crops without harming the quality of the crops.
    (2) Many crops on farms do not receive enough water from rain.
    (3) Most farms use the most efficient irrigation systems available.
    (4) Farm animals can only drink the same water that people do.
    (5) Runoff from farm fields can contain contaminants that make it unusable for drinking.

①②③④⑤

22. Which of the following actions would probably most reduce the amount of water used in the United States?

    (1) encouraging people to drink less water
    (2) building power plants that do not require as much coolant
    (3) increasing the amount of mining that occurs
    (4) consuming more wild-caught fish
    (5) raising fewer cattle on farms

①②③④⑤

23. By some estimates, more than 2,000 gallons of water are required to produce one pound of beef. Only about 700 gallons per pound are accounted for by the water that the cow drinks. What is the most likely source of the remaining 1,300 gallons per pound?

    (1) water used to wash the cow
    (2) water that evaporates from the cow's breath
    (3) water used to process the beef from the cow
    (4) water used to grow the grains that the cow eats
    (5) water that runs into the ground when the cows drink

①②③④⑤

24. Based on the information, which of the following statements is true?

    (1) People in the United States consume most of the world's fresh water for drinking.
    (2) Fish farming and mining account for a majority of the freshwater use in the United States.
    (3) Livestock in the United States drink more water each year than people do.
    (4) Much more fresh water is used in power plants than is used to irrigate crops.
    (5) Industry and agriculture account for the vast majority of freshwater consumption in the United States.

①②③④⑤

GED JOURNEYS

DANICA PATRICK

From an early age, Danica Patrick had a need for speed. Patrick, today one of racing's most popular drivers, first developed an interest in auto racing while driving go-karts as a girl. At age 16, Patrick wanted to pursue a career in racing. She left high school, obtained her GED certificate, and moved to England, where she raced in several open-wheel series. She finished second in England's competitive Formula Ford Festival, the highest finish ever by a woman or an American in the event.

Patrick signed with Rahal Letterman Racing in 2002 and finished third in the Toyota Atlantic Championship two years later. In May 2005, Patrick became only the fourth woman to drive in the Indianapolis 500. She led the race for 19 laps—becoming the first woman to ever lead at Indy—and finished fourth overall in her debut. Later that year at Kansas Speedway, Patrick claimed the top qualifying time and, with it, her first pole position. She was named the Indy Racing League's Rookie of the Year in 2005. Patrick notes that science and technology, ranging from fuel strategies to wind drafts, play large roles in her racing success:

After obtaining her GED certificate, finishing fourth at the Indianapolis 500, and winning her first event, Danica Patrick's career is in high-gear.

> **"Technology provides our pit crews and race strategists with the data they need to make split-second decisions that can make the difference between winning and losing a race."**

In 2008, Patrick became the first female race winner in IndyCar history with a victory in the Indy Japan 300. She has been featured in national magazines, has appeared in various commercials, and has even hosted several television shows.

BIO BLAST: Danica Patrick

- Born in 1982 in Beloit, Wisconsin; raised in Roscoe, Illinois
- Finished fourth at 2005 Indianapolis 500, a record for a female driver
- Captured her first pole position in 2005 at Kansas Speedway
- Published her autobiography *Danica: Crossing the Line* in 2006
- Became the first female driver to win an IndyCar event with a victory at the Indy Japan 300 in 2008

Unit 3: Physical Science

From cooking and baking to developing new technologies and on through unlocking the secrets of the universe, physical science literally guides our every move. Physical science often is divided into chemistry (the study of matter) and physics (the study of the relationship between matter and energy).

Physical science is prominent in the GED Science Test, where it makes up 35 percent of questions. As with the rest of the GED Science Test, physical science questions will assess your ability to successfully interpret passages and visuals. In Unit 3, the introduction of certain skills and the continuation of others, in combination with essential science concepts, will help you prepare for the GED Science Test.

Table of Contents

Lesson 1: Interpret Complex Diagrams 70–71
Lesson 2: Interpret Complex Tables 72–73
Lesson 3: Understand Illustrated Models 74–75
Lesson 4: Interpret Observations 76–77
Lesson 5: Predict Outcomes 78–79
Lesson 6: Use Calculations to Interpret Outcomes 80–81
Lesson 7: Draw Conclusions from Multiple Sources 82–83
Lesson 8: Interpret Multi-Bar and - Line Graphs........ 84–85
Lesson 9: Interpret Pictographs 86–87
Lesson 10: Relate Text and Figures 88–89
Lesson 11: Analyze Results 90–91
Lesson 12: Apply Concepts 92–93
Unit 3 Review 94–99

LESSON 1

Interpret Complex Diagrams

1 Learn the Skill

As you know, **diagrams** are visual aids that show relationships between ideas, objects, or events. Complex diagrams show more sophisticated information than do simple diagrams. When **interpreting complex diagrams,** keep in mind that they may show more than one concept or piece of information.

2 Practice the Skill

By mastering the skill of interpreting complex diagrams, you will improve your study and test-taking skills, especially as they relate to the GED Science Test. Examine the diagram below. Then answer the question that follows.

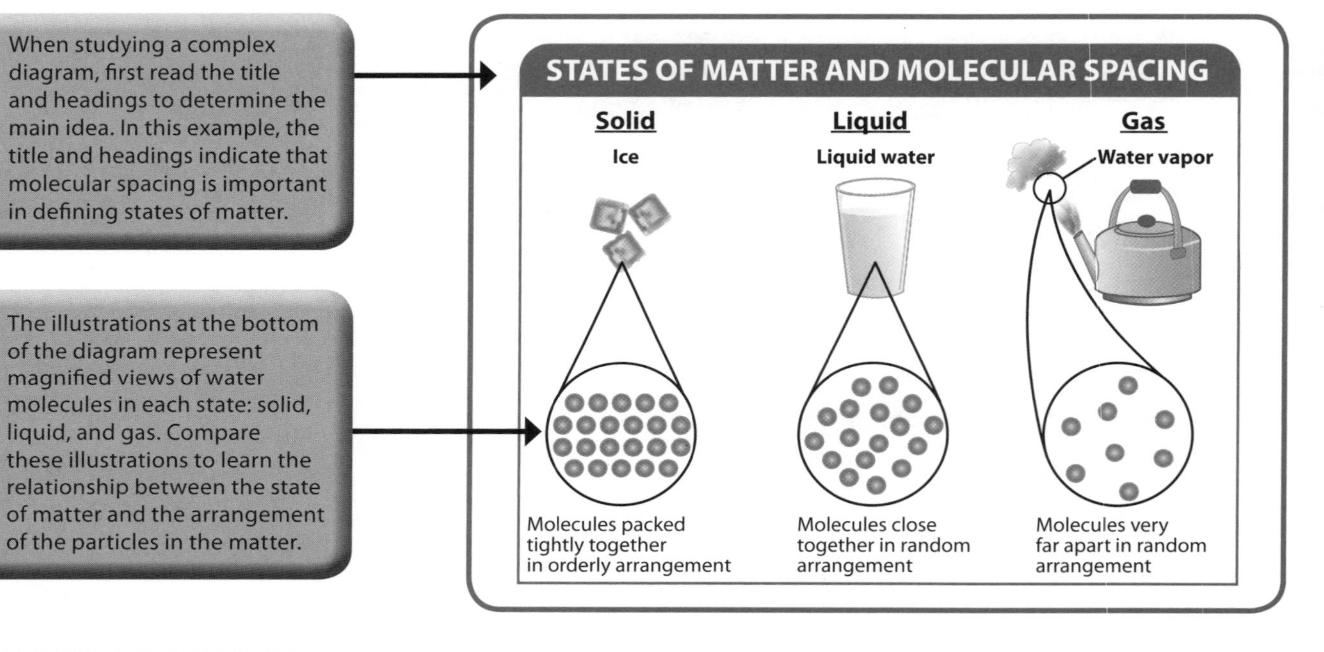

USING LOGIC

Information in a diagram is often arranged in a logical sequence. Therefore, you can use logic to infer the relationships between parts of the diagram. In this diagram, the states of matter are displayed in order of least to greatest spacing between molecules.

1. Based on the diagram, how is the spacing between molecules related to states of matter?

   The spacing between molecules is generally

   (1) smallest in gases
   (2) largest in liquids
   (3) smallest in solids
   (4) constant across all states
   (5) random across all states

3 Apply the Skill

Directions: Choose the one best answer to each question.

Questions 2 and 3 refer to the following paragraph and diagram.

Substances can undergo changes in state. These changes occur when the energy of a system changes. For example, energy must be added to melt a solid into liquid form or to vaporize a liquid into a gas. Energy is released when a gas condenses into a liquid or when a liquid freezes as a solid.

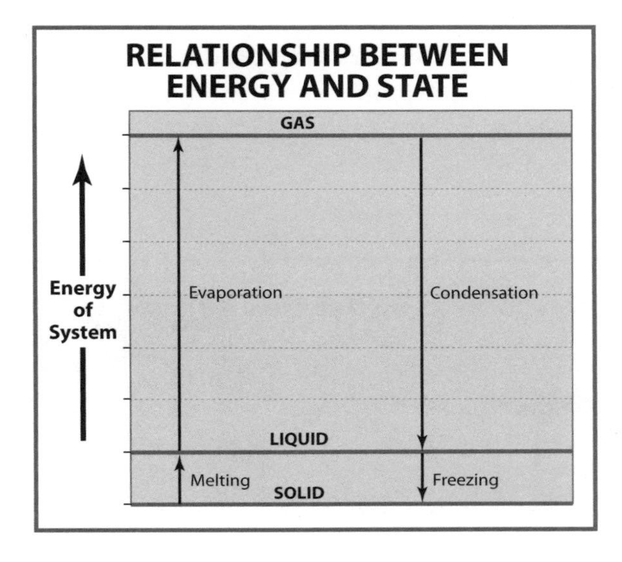

2. Based on the diagram, which is true?

   (1) Removing energy from liquid water changes it to water vapor.
   (2) Adding energy to liquid water changes it to ice.
   (3) The temperature of liquid water does not change as it evaporates.
   (4) Different state changes involve different amounts of energy.
   (5) All state changes involve the same amount of energy.

3. The kilojoule (kJ) is a unit of energy. If a sample of ice requires 6 kJ of energy to melt, how many kilojoules are most likely needed to evaporate the water that is produced?

   (1) about 3 kJ
   (2) about 6 kJ
   (3) about 12 kJ
   (4) about 40 kJ
   (5) about 500 kJ

Questions 4 and 5 refer to the following paragraph and diagram.

A heating curve shows how the temperature of a substance changes as heat is added to it. The melting and boiling points of the substance can be identified from the curve. The diagram shows a heating curve for water.

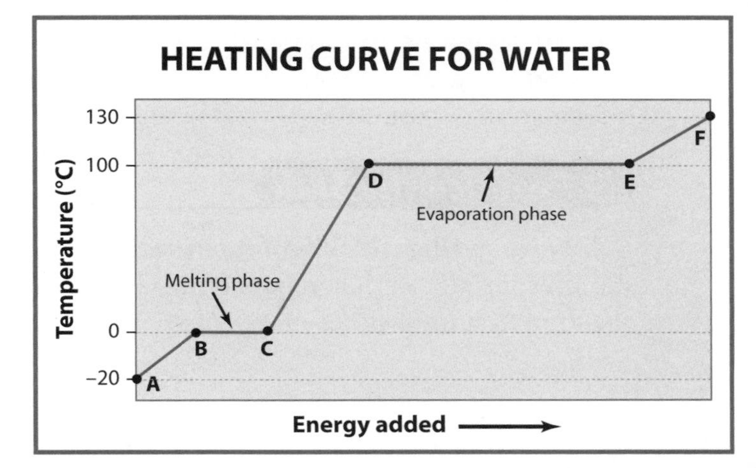

4. Based on the diagram, which of the following statements is true?

   (1) Water continually increases in temperature as it is heated.
   (2) The temperature of a substance changes only during changes of state.
   (3) The temperature of water decreases when the water freezes.
   (4) The addition of energy to a system is always accompanied by an increase in temperature.
   (5) During changes of state, temperature stays constant.

5. Which part of the diagram could tell you the freezing point of water?

   (1) point A
   (2) point B
   (3) the segment between points C and D
   (4) the segment between points D and E
   (5) point F

UNIT 3

LESSON 2

Interpret Complex Tables

1 Learn the Skill

As you learned, a **table** is a graphic tool used to display complex information in an organized and concentrated way. To **interpret complex tables**, you must carefully read the column headings to make sure you know exactly the type of information contained in the table. You also might need to examine information surrounding the table, such as a key or footnotes that provide additional details.

2 Practice the Skill

By mastering the skill of interpreting complex tables, you will improve your study and test-taking skills, especially as they relate to the GED Science Test. Examine the table below. Then answer the question that follows.

UNIT 3

A Typically, all entries in one column provide a particular type of information. Individual row entries are typically related to the entry in the far left column.

B In many cases, important details do not fit into a particular part of the table. Additional symbols may refer you to another place to find the information. Look at the key below the table to find out the meaning of the asterisk.

THE ALKALI METALS

ELEMENT (A)	ATOMIC NUMBER* (B)	ATOMIC WEIGHT** (AMU)
Lithium	3	6.94
Sodium	11	22.99
Potassium	19	39.10
Rubidium	37	85.47
Cesium	55	132.91
Francium	87	223

(B) *atomic number: the number of protons in a single atom of a particular element
**atomic weight: the average mass of atoms of a particular element

MAKING ASSUMPTIONS

In tables that provide numerical data, units are typically stated in the column heading only. All values in the column are expressed in those same units.

1. Based on the table, what trend can you identify?

   The larger the atomic number of an element,

   (1) the smaller its atomic weight
   (2) the greater its atomic weight
   (3) the fewer protons it contains
   (4) the fewer atoms it contains
   (5) the more atoms it contains

3 Apply the Skill

Directions: Choose the one best answer to each question.

Questions 2 through 4 refer to the following paragraph and table.

An ionic compound is made up of particles of two or more elements; one of those particles donates an electron to another particle. A covalent compound is made up of two or more atoms in which atoms share electrons.

BOILING POINTS OF SELECT COMPOUNDS

COMPOUND	FORMULA	TYPE OF COMPOUND	BOILING POINT (°C)
Sodium chloride	NaCl	Ionic	1,413
Hydrogen fluoride	HF	Covalent	20
Hydrogen sulfide	H_2S	Covalent	–61
Calcium iodide	CaI_2	Ionic	1,100
Magnesium fluoride	MgF_2	Ionic	2,239

2. Which compound boils at the highest temperature?

   (1) NaCl
   (2) HF
   (3) H_2S
   (4) CaI_2
   (5) MgF_2

3. A scientist studies an unknown compound and concludes that the compound is ionic. Which of the following is a likely boiling point for the compound?

   (1) –100°C
   (2) 15°C
   (3) 80°C
   (4) 110°C
   (5) 1,050°C

4. Room temperature is about 25°C. Which of the following compounds would you expect to be gases at room temperature?

   (1) NaCl only
   (2) HF and H_2S
   (3) H_2S only
   (4) NaCl, CaI_2, and MgF_2
   (5) HF and MgF_2

Questions 5 and 6 refer to the following paragraph and table.

Six elements listed on the periodic table are known as "noble gases." These elements are very stable and do not naturally form compounds with other elements. Their properties make them important for many commercial uses.

NOBLE GASES AND COMMON USES

ELEMENT	SYMBOL	ATOMIC NUMBER	COMMON USES
Helium	He	2	Balloons, refrigeration*
Neon	Ne	10	Lighting, refrigeration*
Argon	Ar	18	Lighting
Krypton	Kr	36	Lighting
Xenon	Xe	54	Lighting, research
Radon	Rn	86	Medicine

* in liquid form

5. Based on the information, what is the most common commercial use of noble gases?

   (1) refrigeration
   (2) fuel
   (3) medicine
   (4) balloons
   (5) lighting

6. Based on the information, what can you conclude about the noble gas elements?

   Noble gases

   (1) do not occur naturally
   (2) can exist as liquids as well as gases
   (3) are the most common elements
   (4) have large atomic numbers
   (5) are extremely heavy elements

UNIT 3

LESSON 3

Understand Illustrated Models

1 Learn the Skill

An **illustrated model** can be used to represent objects that are too large or too small to be shown in actual size. By **understanding illustrated models**, you can interpret processes that occur too quickly or too slowly to be directly observed. An illustrated model can be expressed in two-dimensional form, in three-dimensional form, or as a mathematical equation.

2 Practice the Skill

By mastering the skill of understanding illustrated models, you will improve your study and test-taking skills, especially as they relate to the GED Science Test. Examine the model below. Then answer the question that follows.

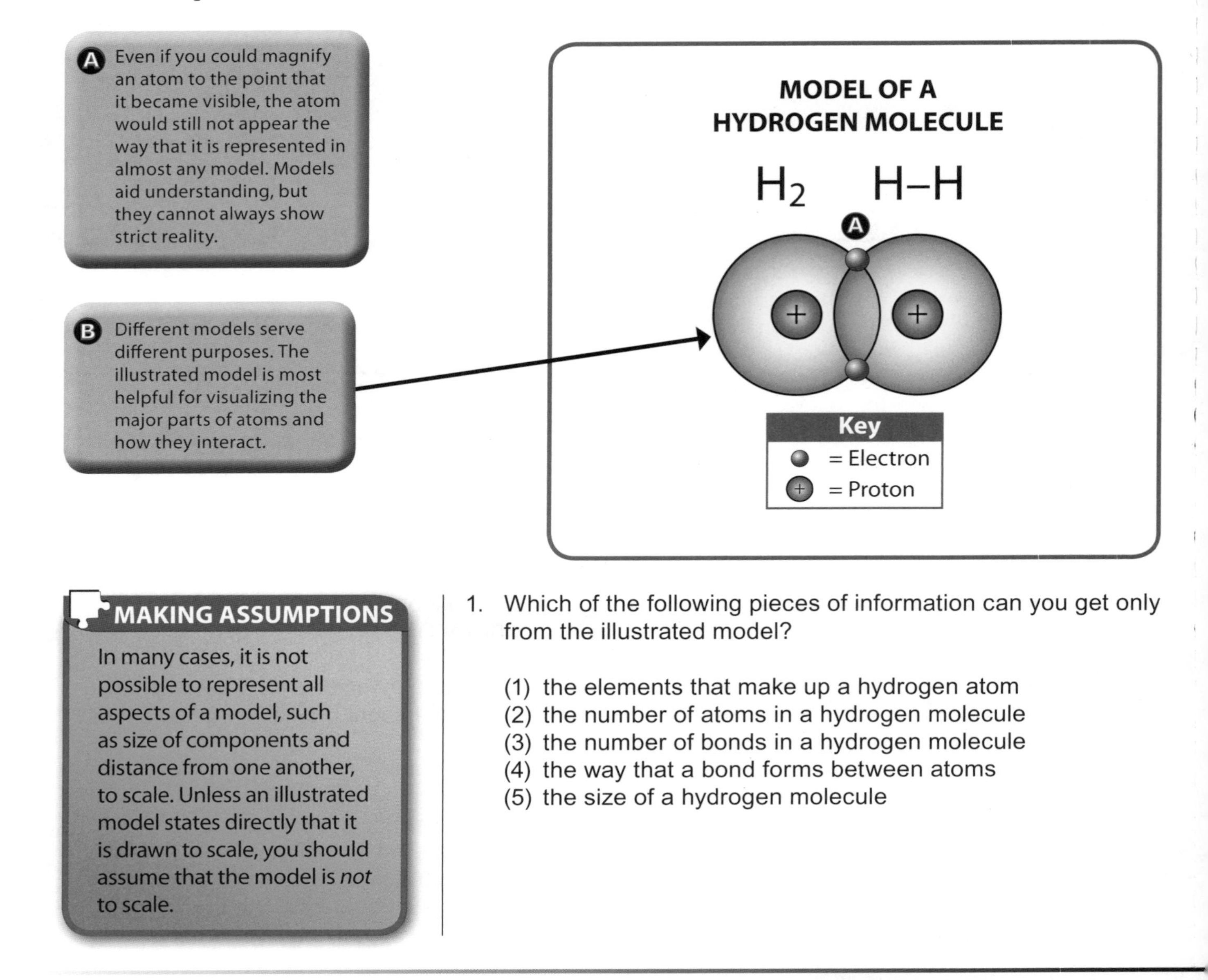

A Even if you could magnify an atom to the point that it became visible, the atom would still not appear the way that it is represented in almost any model. Models aid understanding, but they cannot always show strict reality.

B Different models serve different purposes. The illustrated model is most helpful for visualizing the major parts of atoms and how they interact.

MAKING ASSUMPTIONS

In many cases, it is not possible to represent all aspects of a model, such as size of components and distance from one another, to scale. Unless an illustrated model states directly that it is drawn to scale, you should assume that the model is *not* to scale.

1. Which of the following pieces of information can you get only from the illustrated model?

   (1) the elements that make up a hydrogen atom
   (2) the number of atoms in a hydrogen molecule
   (3) the number of bonds in a hydrogen molecule
   (4) the way that a bond forms between atoms
   (5) the size of a hydrogen molecule

3 Apply the Skill

Directions: Choose the one best answer to each question.

Questions 2 through 4 refer to the following paragraph and illustrated model.

Hydrogen is the simplest of the elements, with each atom consisting of a single proton and a single electron. Its name comes from the Greek words for "that which forms water." The electron moves so quickly around the proton that it cannot be seen. Although the location of the electron is represented in the model by a ring for simplicity, a "cloud" might represent the location of the electron more accurately.

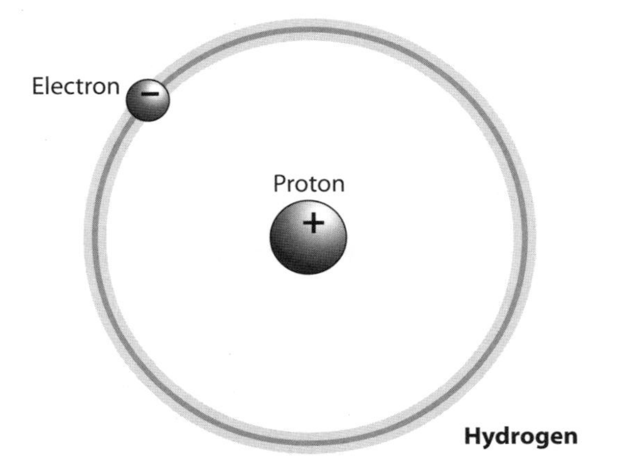

2. What does the large ring surrounding the "+" particle indicate?

   (1) the location of the "–" particle
   (2) the location of the "+" particle
   (3) the structure of a hydrogen molecule
   (4) the location of the atom's positron
   (5) the location of the molecule's positron

3. The hydrogen atom is too small to be shown in actual size. What is another reason that the atom is shown as an illustrated model?

   (1) The electron is much larger than the nucleus.
   (2) The electron moves too quickly to be seen directly.
   (3) Hydrogen atoms are too rare to be seen easily.
   (4) The positive charge of the proton makes it hard to see.
   (5) The electron is too close to the nucleus to tell one from the other.

4. How could you change this two-dimensional model into a three-dimensional model?

   (1) add another electron to the model
   (2) use two balls and a long wire to recreate the model
   (3) draw an arrow on the model showing the direction of movement
   (4) display the model on posterboard
   (5) add two more "+" symbols to the model

Questions 5 and 6 refer to the following information and models.

Each of the models below represents a hydrogen molecule.

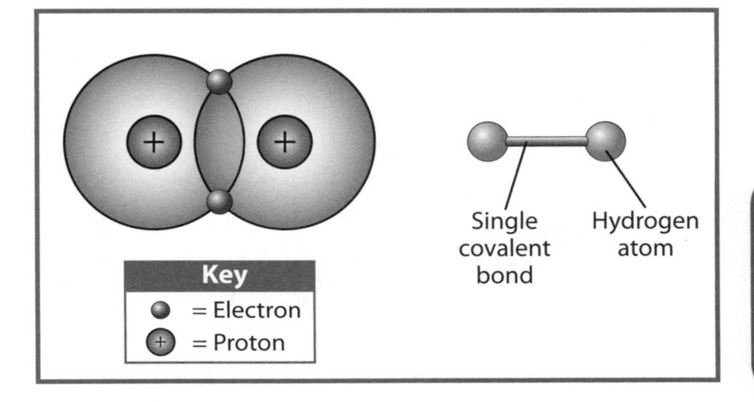

5. Based on the models, which of the following is the best description of a covalent bond?

   (1) a stick-like connection between two atoms
   (2) a proton shared by two atoms
   (3) a set of electrons shared between atoms
   (4) an overlap of the circular borders of two atoms
   (5) an overlap of a proton and an electron within an atom

6. For which of the following purposes would the ball-and-stick model be most useful?

   The ball-and-stick model would be most useful for showing the

   (1) relative positions of atoms in a molecule
   (2) overlap between atoms in a true bond
   (3) size of a molecule
   (4) relative sizes of electrons and protons
   (5) actual appearance of each atom

UNIT 3

Interpret Observations

1 Learn the Skill

Observations are pieces of information gathered with the senses. To **interpret observations** means to figure out what those observations mean. Understanding the meaning of observations is an important part of the scientific process.

2 Practice the Skill

By mastering the skill of interpreting observations, you will improve your study and test-taking skills, especially as they relate to the GED Science Test. Examine the paragraph and diagram below. Then answer the question that follows.

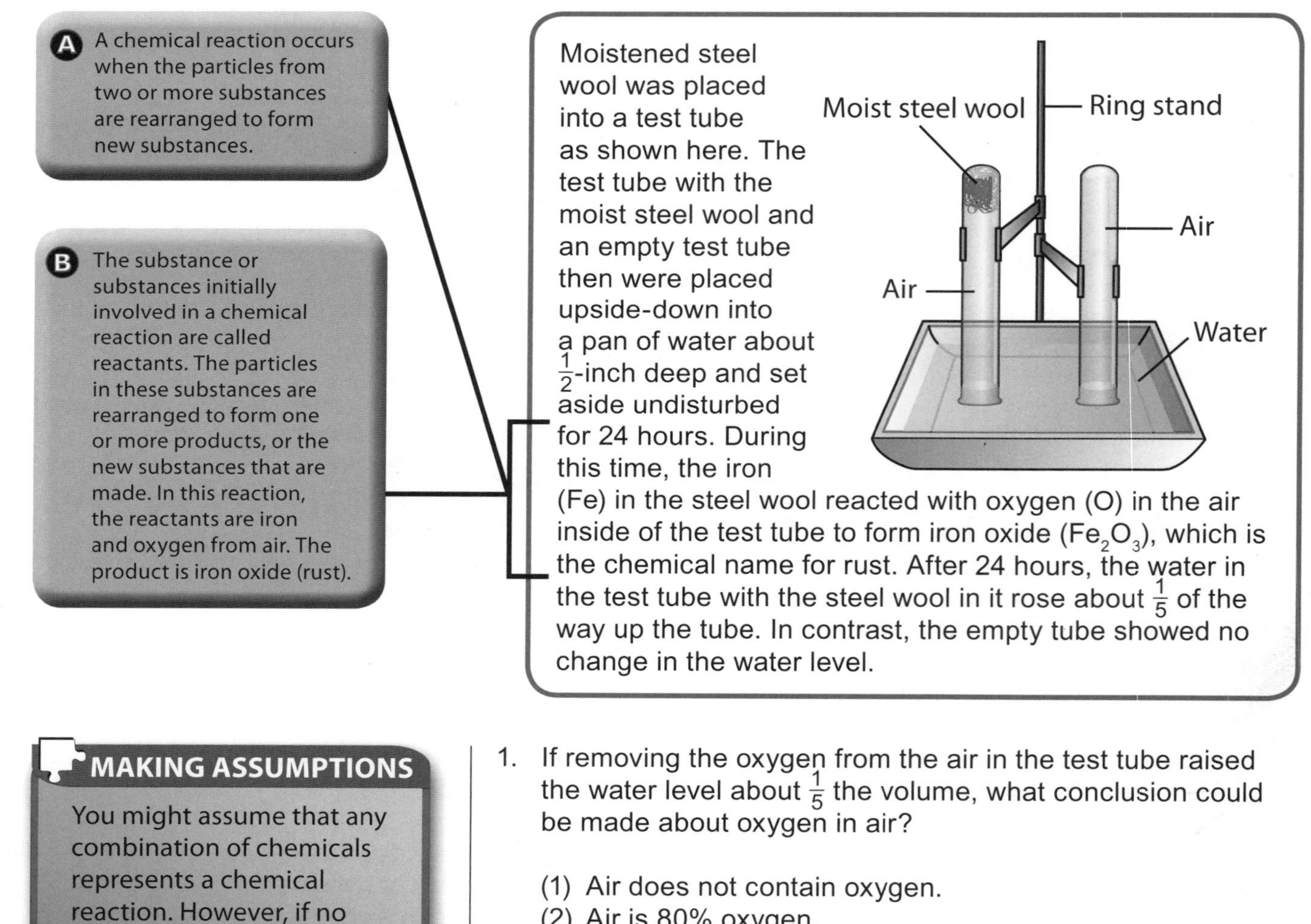

A A chemical reaction occurs when the particles from two or more substances are rearranged to form new substances.

B The substance or substances initially involved in a chemical reaction are called reactants. The particles in these substances are rearranged to form one or more products, or the new substances that are made. In this reaction, the reactants are iron and oxygen from air. The product is iron oxide (rust).

Moistened steel wool was placed into a test tube as shown here. The test tube with the moist steel wool and an empty test tube then were placed upside-down into a pan of water about $\frac{1}{2}$-inch deep and set aside undisturbed for 24 hours. During this time, the iron (Fe) in the steel wool reacted with oxygen (O) in the air inside of the test tube to form iron oxide (Fe_2O_3), which is the chemical name for rust. After 24 hours, the water in the test tube with the steel wool in it rose about $\frac{1}{5}$ of the way up the tube. In contrast, the empty tube showed no change in the water level.

MAKING ASSUMPTIONS

You might assume that any combination of chemicals represents a chemical reaction. However, if no new substances can be observed, then a chemical reaction has not occured.

1. If removing the oxygen from the air in the test tube raised the water level about $\frac{1}{5}$ the volume, what conclusion could be made about oxygen in air?

(1) Air does not contain oxygen.
(2) Air is 80% oxygen.
(3) Oxygen is about $\frac{1}{5}$ the volume of air.
(4) Air contains trace amounts of oxygen.
(5) Oxygen is the main component of air.

3 Apply the Skill

Directions: Choose the one best answer to each question.

Questions 2 and 3 refer to the following information and illustration.

When a metal reacts with an acid, a salt and hydrogen gas are formed. The general word equation for this reaction is:

Metal + acid ⟶ salt + hydrogen

The activity series is often used to predict whether certain reactions will occur. More reactive metals appear at the top of the list, while less reactive metals appear at the bottom. Metals listed below hydrogen on the activity series do not react with acids to produce hydrogen gas. Three small pieces of three different metals were placed in separate test tubes as shown here. Acid was added to each test tube.

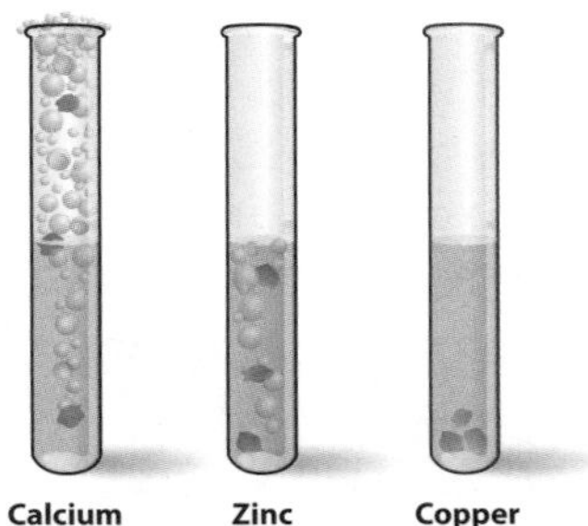

2. Which of the following correctly lists the metals in order of decreasing reactivity?

   (1) copper, zinc, calcium
   (2) calcium, copper, zinc
   (3) copper, calcium, zinc
   (4) calcium, zinc, copper
   (5) zinc, copper, calcium

3. Based on the information and illustration, which of the following statements is true?

   (1) Copper is listed above hygrogen on the activity series.
   (2) Zinc is listed below hydrogen on the activity series
   (3) Both zinc and calcium are listed above hydrogen on the activity series.
   (4) Calcium is listed below hydrogen on the activity series.
   (5) Calcium, copper, and zinc are listed below hydrogen on the activity series.

Questions 4 and 5 refer to the following paragraph and table.

Chemical change requires a chemical reaction. In a chemical reaction, the particles of subtances are rearranged to form new substances. Physical changes are those changes that do not result in a new substance being formed.

CHEMICAL AND PHYSICAL CHANGES		
MATERIAL	CHANGE	OBSERVATION
Candle	Melted	Solid candle became liquid wax.
Candle	Burned	Candle seemed to disappear.
Silver	Melted	Solid silver became liquid silver.
Silver	Tarnished	Silver developed a dark-colored coating on it.
Paper	Torn	A large piece of paper became smaller pieces.
Paper	Burned	Paper became ashes and smoke.

4. Based on the information, which of the following statements is true?

   (1) Melting silver is an example of a chemical change.
   (2) Burning paper is an example of a chemical change.
   (3) Tearing paper is an example of a chemical change.
   (4) Silver tarnishes as a result of a physical change.
   (5) Candle wax melts as a result of a chemical change.

5. Based on the information, which of the following is the best indication that when silver tarnishes, a chemical reaction takes place?

   (1) Solid silver becomes liquid silver.
   (2) No new substances are formed.
   (3) The color of the silver remains the same.
   (4) The dark coating is a new substance.
   (5) When silver tarnishes, it becomes lighter.

UNIT 3

LESSON 5

Predict Outcomes

1 Learn the Skill

Scientists follow the general sequence of 1) observation, 2) analysis, and 3) **predicting outcomes** as they carry out experiments to discover new knowledge about the universe. By looking for patterns in data and observations, they are able to develop hypotheses that can be tested through further experimentation.

2 Practice the Skill

By mastering the skill of predicting outcomes, you will improve your study and test-taking skills, especially as they relate to the GED Science Test. Examine the paragraph and table below. Then answer the question that follows.

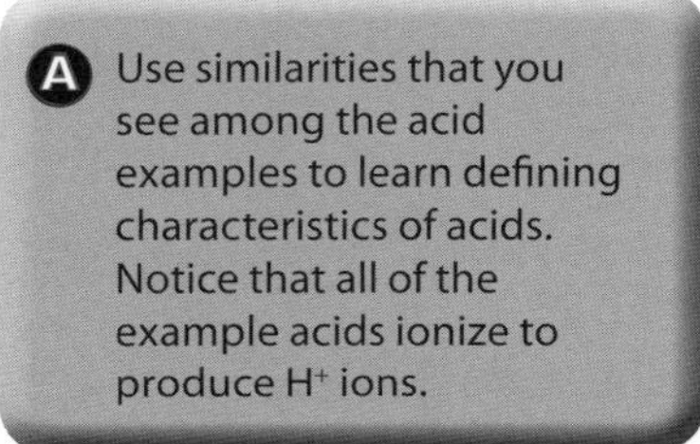

B Use similarities that you see among the base examples to learn defining characteristics of bases. Notice that all of the example bases ionize to produce OH^- ions.

We encounter acids, bases, and salts frequently in the world around us. Acids, bases, and salts are similar in that they have molecular compositions: two or more types of atoms bond together to form them. In addition, all three ionize when they are dissolved in water. This means that the parent molecules break into charged particles called ions. Several examples are shown below.

ACIDS, BASES, AND SALTS

	EXAMPLE	IONIZATION REACTION
Acids	HCl HBr HI	$HCl \longrightarrow H^+ + Cl^-$ $HBr \longrightarrow H^+ + Br^-$ $HI \longrightarrow H^+ + I^-$
Bases	NaOH KOH LiOH	$NaOH \longrightarrow Na^+ + OH^-$ $KOH \longrightarrow K^+ + OH^-$ $LiOH \longrightarrow Li^+ + OH^-$
Salts	NaCl LiBr KI	$NaCl \longrightarrow Na^+ + Cl^-$ $LiBr \longrightarrow Li^+ + Br^-$ $KI \longrightarrow K^+ + I^-$

USING LOGIC

Looking for patterns and trends in groups of data or observations can help you to make generalizations. The ability to generalize about a concept can make it easier to predict future outcomes.

1. Based on the patterns you see in the table above, which set of compounds would have the most properties in common with the compound HF?

   (1) LiOH, KOH, HI
   (2) NaCl, KI, LiOH
   (3) HBr, HI, HCl
   (4) NaOH, LiOH, KOH
   (5) LiBr, KI, NaCl

3 Apply the Skill

Directions: Choose the one best answer to each question.

Questions 2 through 6 refer to the following paragraphs and equations.

Acids are compounds that ionize to form H^+ (hydrogen ions). The concentration of H^+ in solution is used to measure the acidity of a solution. The greater the H^+ concentration, the more acidic the solution is. HCl and H_2SO_4 are examples of acids. Their ionization reactions are shown below.

$$HCl \longrightarrow H^+ + Cl^-$$
$$H_2SO_4 \longrightarrow 2H^+ + SO_4^{2-}$$

Bases are compounds that can react with H^+ ions to reduce their concentration in solution. Bases can therefore neutralize acidic solutions. An example of a base is NaOH. When NaOH is mixed with an acid, OH^- (hydroxide ion) combines with H^+ from the acid to form water. The overall equation showing the reaction of NaOH with HCl is shown below. A general reaction scheme is shown underneath.

$$HCl + NaOH \longrightarrow H_2O + NaCl$$
$$\textbf{Acid} + \textbf{Base} \longrightarrow \textbf{Water} + \textbf{Salt}$$

Notice that water is one product of the reaction and salt is another product. These reactions are called neutralization reactions because both H^+ and OH^- ion concentrations are lowered as a result of the reaction.

2. A student mixes the following pairs of compounds in solution. What is predicted to occur in each case?

KOH + NaOH
HI + KBr
H_2SO_4 + KOH

(1) no reaction, reaction, reaction
(2) reaction, reaction, reaction
(3) no reaction, reaction, no reaction
(4) reaction, no reaction, no reaction
(5) no reaction, no reaction, reaction

3. Using the definition of acidity given in the text, what would you predict to be the relative acidity of two solutions prepared using equal numbers of molecules of HCl and H_2SO_4?

The HCl solution would be

(1) exactly as acidic as the H_2SO_4 solution
(2) twice as acidic as the H_2SO_4 solution
(3) half as acidic as the H_2SO_4 solution
(4) three times as acidic as the H_2SO_4 solution
(5) one-third as acidic as the H_2SO_4 solution

4. Water and a salt are formed when HCl (an acid) is mixed with LiOH (a base).

The salt that forms is shown as

(1) LiH
(2) HOH
(3) LiHCl
(4) OHCl
(5) LiCl

5. Which reaction correctly represents the reaction of hydrocyanic acid (HCN) with potassium hydroxide (KOH)?

(1) $HCN + H_2O \longrightarrow KOH + NaCl$
(2) $HCN + KOH \longrightarrow HK + CNOH$
(3) $HCN + KOH \longrightarrow KCN + H_2O$
(4) $KOH + H_2O \longrightarrow HCN + KCN$
(5) $KOH + HCN \longrightarrow KON + CH_2$

6. Potassium bromide (KBr) is a salt. It could be produced by which of the following reactions?

(1) HBr + KOH
(2) HF + NaI
(3) HCl + NaOH
(4) H_2SO_4 + KOH
(5) HBr + KI

UNIT 3

Use Calculations to Interpret Outcomes

1 Learn the Skill

Scientific texts often include diagrams with numerical values listed on them. The diagrams may describe an event that occurred in an experiment or a test. To help make the numbers more meaningful, you could include them in a calculation of a rate or force that was tested in the experiment. When you use this process, you are **using calculations to interpret outcomes**.

2 Practice the Skill

By mastering the skill of using calculations to interpret outcomes, you will improve your study and test-taking skills, especially as they relate to the GED Science Test. Examine the text and diagram below. Then answer the question that follows.

A This text provides information related to a calculation. It states that to calculate speed, you divide time into distance, or $s = \frac{d}{t}$.

B This question asks you to use a calculation to interpret an outcome. To calculate speed, use the formula $s = \frac{d}{t}$. Use data from the diagram to find numbers for distance and time.

Speed is a measure of how fast an object moves, and it is calculated by dividing the distance that the object moved by the time required for the movement. Displacement describes how far an object travels, or its total change in position.

A commuter uses a scooter to travel 4 miles to work each day. The commuter's route is a direct route to the hospital where she works and then a direct route home. The chart below shows her average speed during a one-way trip to work or a one-way trip home.

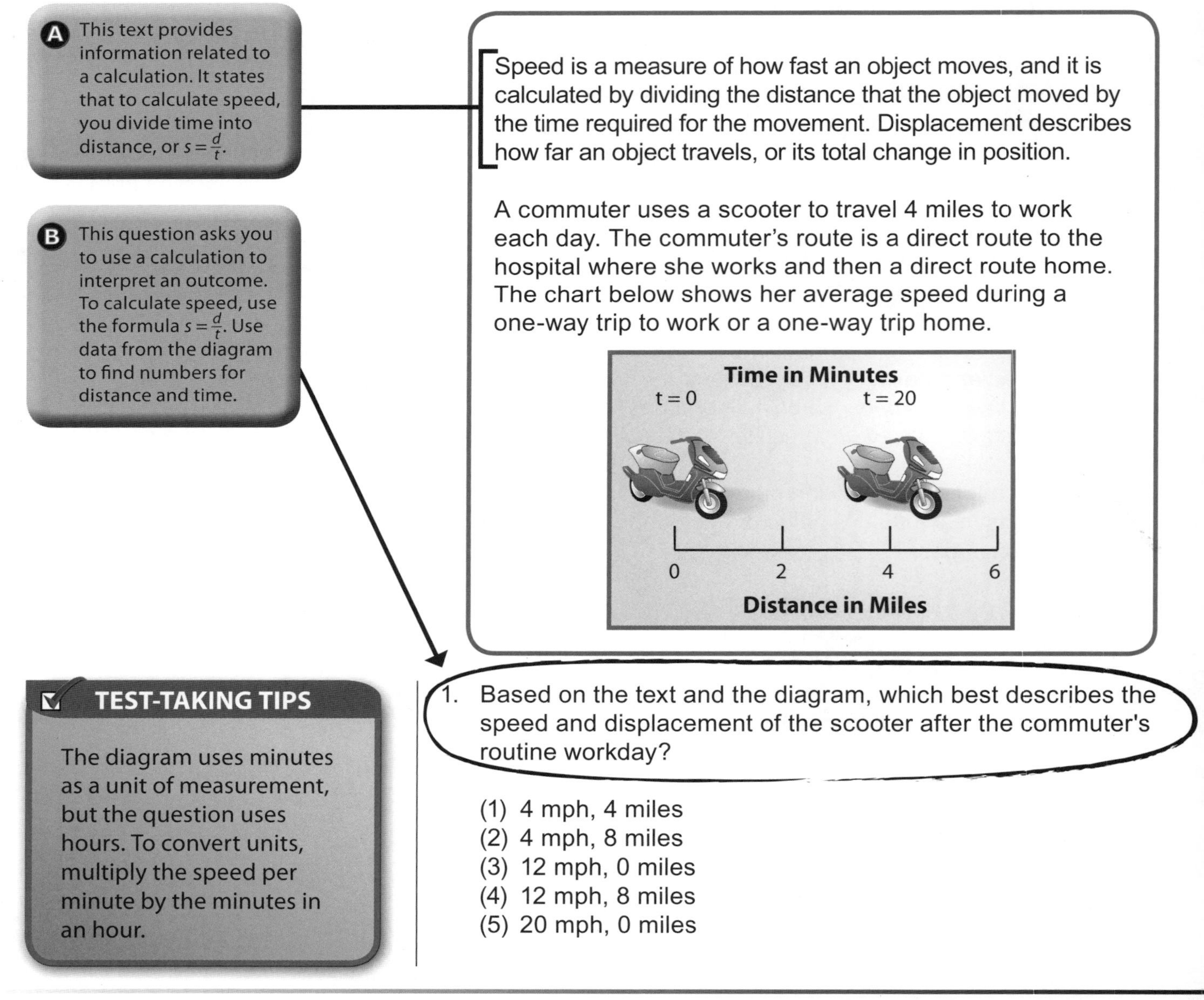

TEST-TAKING TIPS

The diagram uses minutes as a unit of measurement, but the question uses hours. To convert units, multiply the speed per minute by the minutes in an hour.

1. Based on the text and the diagram, which best describes the speed and displacement of the scooter after the commuter's routine workday?

(1) 4 mph, 4 miles
(2) 4 mph, 8 miles
(3) 12 mph, 0 miles
(4) 12 mph, 8 miles
(5) 20 mph, 0 miles

③ Apply the Skill

Directions: Choose the one best answer to each question.

Questions 2 and 3 refer to the following paragraph and diagram.

Newton's second law states that the net force on an object equals its mass multiplied by its acceleration. This law is stated in the equation $F = ma$, where F stands for the net force (expressed in newtons), m stands for the object's mass, and a stands for the object's acceleration.

3 kg

5 m/s^2

F

2. How much force is being applied to the book in the diagram?

   (1) 0.6 newtons
   (2) 3 newtons
   (3) 5 newtons
   (4) 8 newtons
   (5) 15 newtons

3. If a person applied the same force to move more books, what effect would it have on acceleration?

   Acceleration would

   (1) not change
   (2) increase
   (3) decrease
   (4) initially decrease and then gradually increase
   (5) initially increase and then decrease

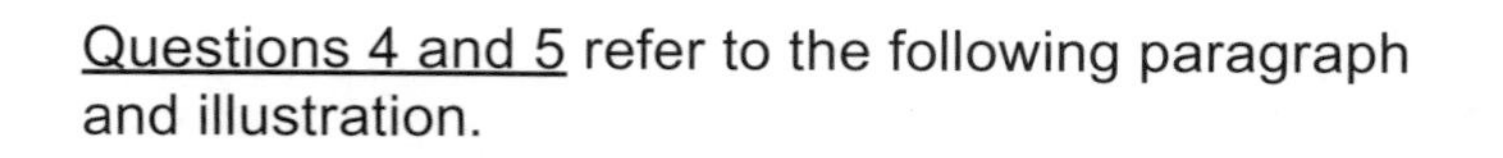
Questions 4 and 5 refer to the following paragraph and illustration.

Newton's third law of motion states that for every force in nature, there is an equal and opposite force. For example, when a person stands on a floor, the person's weight pushes down on the floor. To hold the weight, the floor pushes up on the person's feet with an equal and opposite force. In the illustration below, the person has a mass of 50 kg. The force he exerts on the floor is equal to his weight. Near Earth's surface, an object's weight in newtons is equal to its mass in kilograms times 9.8. Therefore, the person's weight is 50 kg × 9.8 = 490 N.

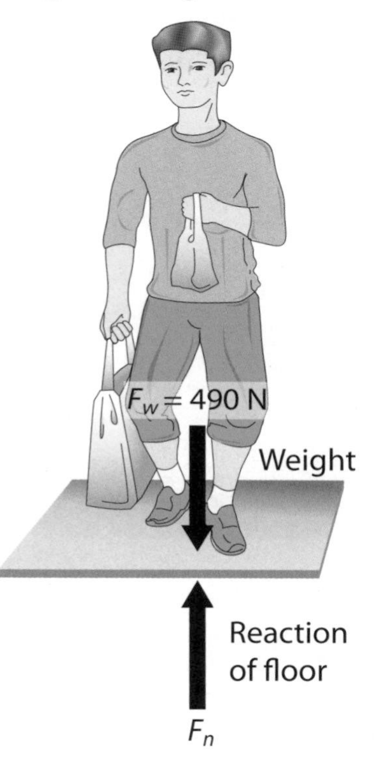

4. What is the value of F_n?

   (1) 9.8 newtons
   (2) 19.6 newtons
   (3) 50 newtons
   (4) 490 newtons
   (5) 24,500 newtons

5. What is the value of F_n for a person who has a mass of 65 kg?

   (1) 6.6 newtons
   (2) 9.8 newtons
   (3) 65 newtons
   (4) 637 newtons
   (5) 6,242.6 newtons

UNIT 3

LESSON 7

Draw Conclusions from Multiple Sources

1 Learn the Skill

You make an inference by using facts, evidence, experience, and reasoning to make an educated guess. Then you can draw a **conclusion**, or an explanation or a judgment usually based on inferences. You can **draw conclusions from multiple sources** by using the information in charts, diagrams, tables, and text to make a statement that explains all of your observations and the facts that are presented.

2 Practice the Skill

By mastering the skill of drawing conclusions from multiple sources, you will improve your study and test-taking skills, especially as they relate to the GED Science Test. Examine the paragraph, illustration, and strategies below. Then answer the question that follows.

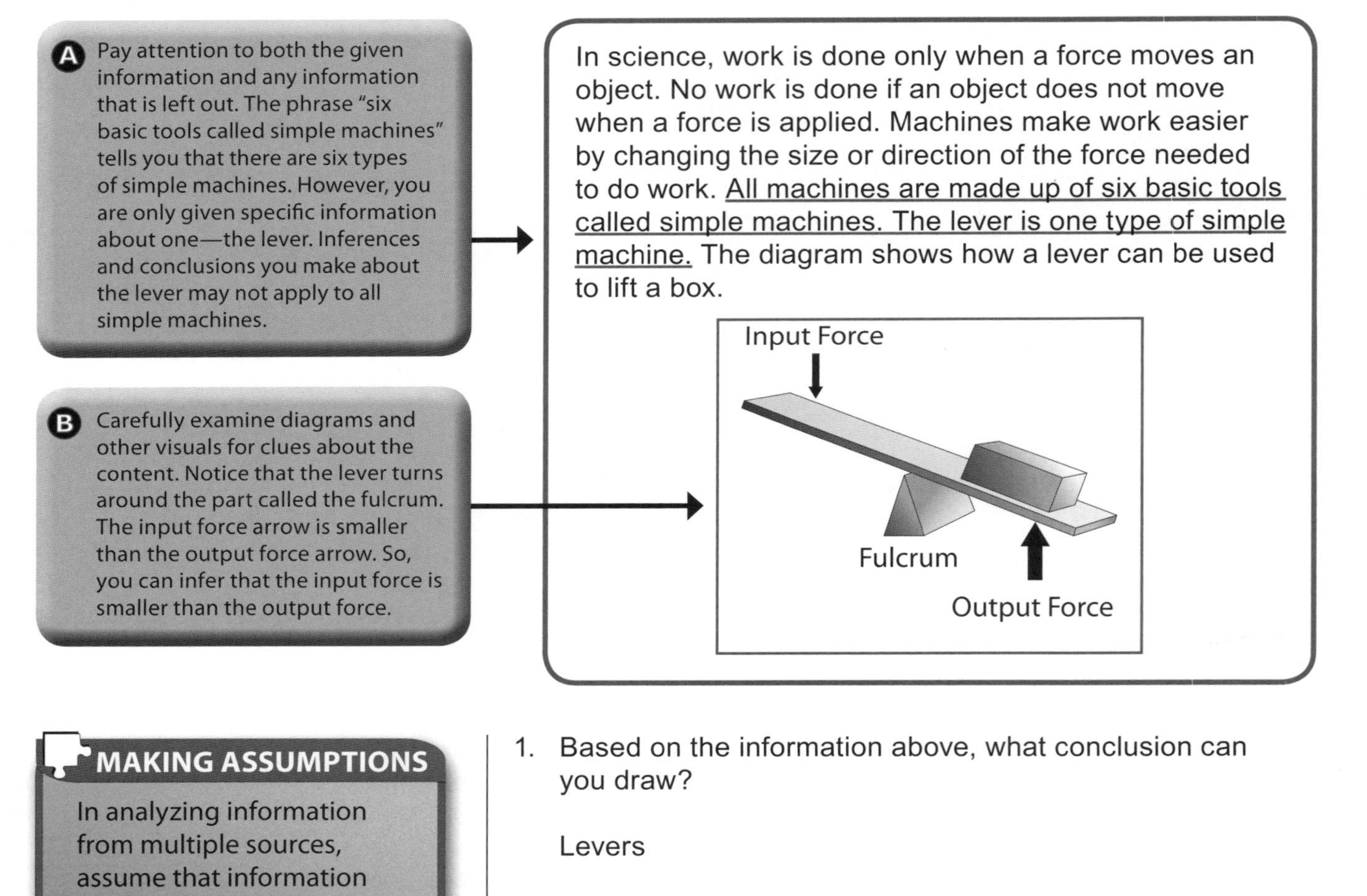

A Pay attention to both the given information and any information that is left out. The phrase "six basic tools called simple machines" tells you that there are six types of simple machines. However, you are only given specific information about one—the lever. Inferences and conclusions you make about the lever may not apply to all simple machines.

B Carefully examine diagrams and other visuals for clues about the content. Notice that the lever turns around the part called the fulcrum. The input force arrow is smaller than the output force arrow. So, you can infer that the input force is smaller than the output force.

In science, work is done only when a force moves an object. No work is done if an object does not move when a force is applied. Machines make work easier by changing the size or direction of the force needed to do work. All machines are made up of six basic tools called simple machines. The lever is one type of simple machine. The diagram shows how a lever can be used to lift a box.

MAKING ASSUMPTIONS

In analyzing information from multiple sources, assume that information from one source complements information from another source. For example, diagrams can help show how a topic discussed in the text works.

1. Based on the information above, what conclusion can you draw?

   Levers

   (1) are the only simple machines that can increase the size of the force
   (2) make work easier by changing the direction of the applied force
   (3) do not do work
   (4) are the only type of simple machine
   (5) apply force to the fulcrum

③ Apply the Skill

Directions: Choose the one best answer to each question.

Questions 2 through 5 refer to the following paragraph, table, and diagram.

A pulley is a simple machine that is used to lift objects. A pulley has two main parts—a grooved wheel and a rope. The rope fits into the groove of the wheel. Pulleys may change the direction of the applied force or the amount of applied force required to lift an object. Pulleys do not change the amount of work that is done. The size of applied force needed to lift an object with a pulley is less because the force is applied over a longer distance. The table below shows the advantages of the three main types of pulleys. A block-and-tackle pulley contains at least one fixed pulley and one movable pulley, as the diagram shows.

TYPE OF PULLEY	CHANGE IN DIRECTION OF FORCE	CHANGE IN SIZE OF FORCE
Single, fixed	Yes	No
Movable	No	Yes
Block and tackle	Yes	Yes

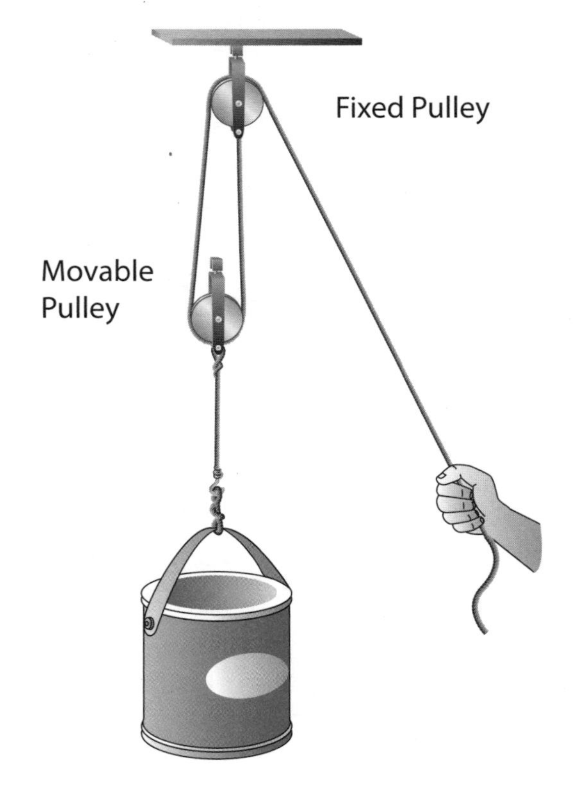

2. Based on the information, what can you conclude about using a single, fixed pulley?

   (1) Less force is applied over a longer distance.
   (2) Single, fixed pulleys are only used as a part of the block and tackle.
   (3) A force of 10 N must be applied to the pulley to lift an object that weighs 10 N.
   (4) Pulling up on the rope lifts the object.
   (5) No work is done when the object is lifted.

3. Based on the information in the diagram and the table, what conclusion can you make about the user of a block and tackle?

   The user of a block and tackle

   (1) pulls down on the rope to make the object move down
   (2) uses less force to lift an object than would be needed without the pulley
   (3) pulls the rope a shorter distance than the object will move
   (4) does not use any force to lift the object
   (5) does more work to lift the object with the pulley than without the pulley

4. Based on the information above, what conclusion can you draw about the benefit of pulleys?

   (1) They always decrease the size of the applied force that is needed.
   (2) Block-and-tackle pulleys can be combined to make movable pulleys.
   (3) They never change the direction of the applied force.
   (4) Some pulleys allow you to use your own weight to pull an object upward.
   (5) All pulleys allow you to do less work when you lift an object.

5. Suppose 30 joules (J) of work is done to lift the paint can without a machine. Based on the information, how much work is done when using a single pulley to lift the same can?

   (1) 30 J
   (2) 45 J
   (3) 90 J
   (4) 120 J
   (5) 180 J

LESSON 8

Interpret Multi-Bar and -Line Graphs

1 Learn the Skill

As you have learned, graphs present data in a visual way. Line graphs can be used to show change over time. Bar graphs can be used to show the relationship of various items. Knowing how to **interpret multi-bar and -line graphs** will enable you to answer questions about information contained in graphs.

2 Practice the Skill

By mastering the skill of interpreting multi-bar and -line graphs, you will improve your study and test-taking skills, especially as they relate to the GED Science Test. Study the information and strategies below. Then answer the question that follows.

A The axis labels tell you which variables are compared in the graph. The *y*-axis shows the amount of energy consumed. The *x*-axis show the years in which it was consumed. Note the range of data shown on the axes. This graph only contains information from 2005 and 2006.

B Multi-bar and -line graphs usually contain a legend. The legend in this bar graph tells you that electrical energy and gas energy are shown on the graph. The graph does not provide information about other types of energy.

The energy that people use in their homes comes from several sources. Most appliances use electrical energy. Many homes are heated by burning natural gas. The bar graph shows the amount of electrical and gas energy used by a community in two consecutive years.

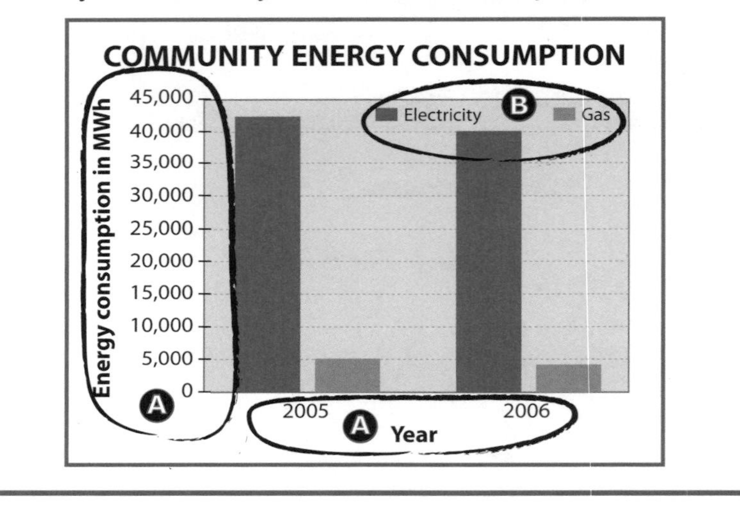

MAKING ASSUMPTIONS

Avoid making assumptions that a graph cannot support. For example, you might assume that people use more gas energy than wind energy. However, the graph does not support that assumption.

1. Based on the information in the graph, what can you conclude about the energy use of this community?

   (1) People used less solar energy than electrical energy.
   (2) People used less gas energy in 2007 than in 2006.
   (3) People used more electrical energy in 2006 than in 2005.
   (4) People used less gas energy in 2006 than in 2005.
   (5) People used more gas energy than electrical energy.

UNIT 3

3 Apply the Skill

Directions: Choose the one best answer to each question.

Question 2 refers to the following paragraph and graph.

Many scientists are concerned that an excessive amount of carbon dioxide (CO_2) in Earth's atmosphere is contributing to climate change. Carbon dioxide is produced when fossil fuels are burned in order to change their stored chemical energy into a form of energy that people can use. The graph below shows the projected carbon dioxide emissions for both developed countries and developing countries.

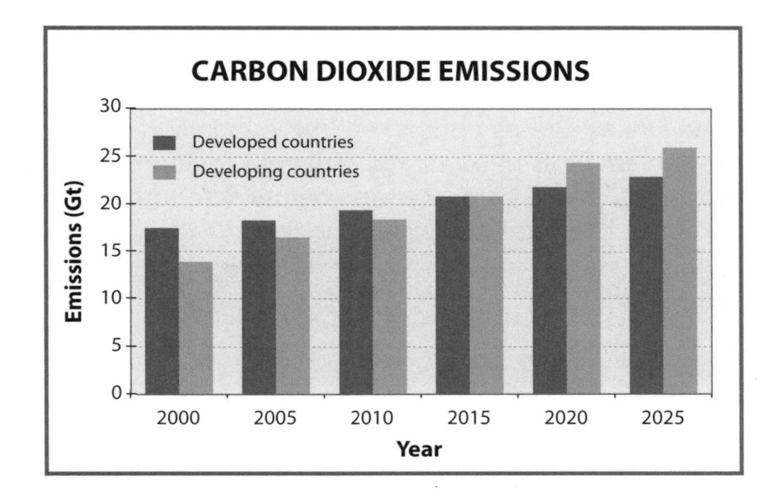

2. Which of the following statements is best supported by the graph?

   The amount of CO_2 released by developing countries

   (1) is currently greater than the amount released by developed countries
   (2) is projected to decrease
   (3) will be equal to the amount released by developed countries in 2015
   (4) does not contribute to climate change
   (5) is expected to increase much more slowly than the amount released by developed countries

Questions 3 and 4 refer to the following graph.

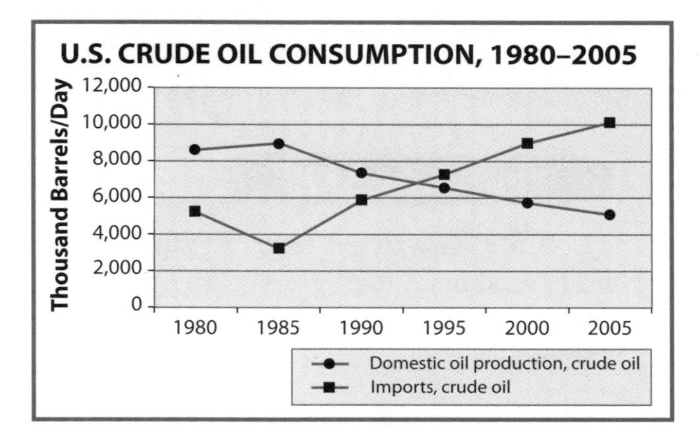

3. What information does the two-line graph above show that a single-line graph could not show?

   (1) how oil consumption varies over time
   (2) the daily amount of oil consumed each year
   (3) the trend in oil consumption from 1980 to 2005
   (4) the number of imported barrels of oil consumed each day from 1980 to 2005
   (5) a comparison between the consumption of domestic oil and the consumption of imported oil

4. Based on the information in the graph, which of the following statements is accurate?

   (1) Domestic oil production increased during the period shown.
   (2) Domestic oil production did not change during the period shown.
   (3) During the period shown, domestic oil production increased and oil imports decreased.
   (4) During the period shown, domestic oil production decreased and oil imports increased.
   (5) Domestic oil production and oil imports both increased during the period shown.

LESSON 9

Interpret Pictographs

1 Learn the Skill

A **pictograph** is a type of graph or chart that uses symbols instead of lines or bars to represent numerical data. Pictographs are not as precise as graphs. For some applications, however, pictographs are more appropriate than graphs. Knowing how to **interpret pictographs** can enable you to quickly compare several groups of data at once, making it easy to see general trends.

2 Practice the Skill

By mastering the skill of interpreting pictographs, you will improve your study and test-taking skills, especially as they relate to the GED Science Test. Examine the paragraph and pictograph below. Then answer the question that follows.

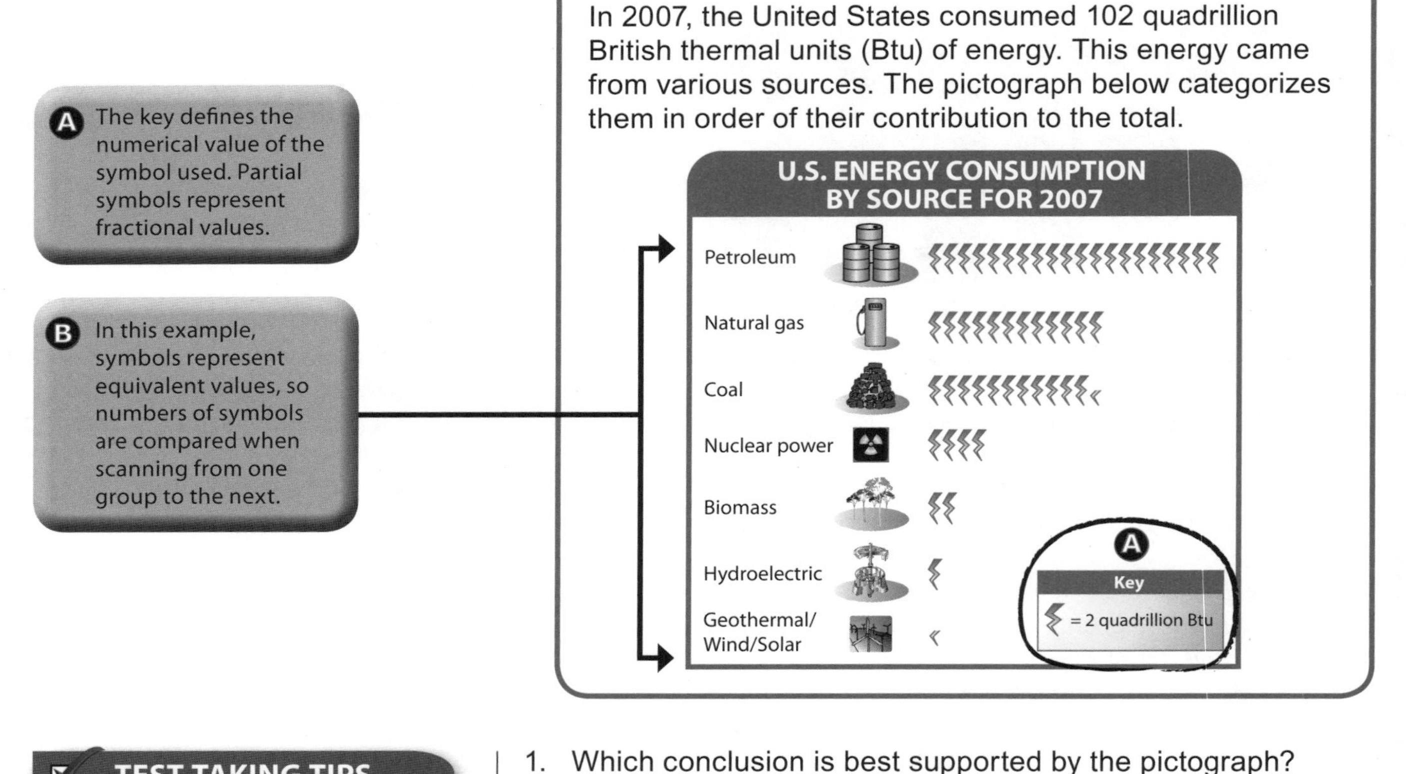

TEST-TAKING TIPS

Scan the rows of symbols to get a general sense of how the numbers vary from row to row. This will allow you to estimate differences between rows.

1. Which conclusion is best supported by the pictograph?

   (1) Coal meets most U.S. energy needs.
   (2) Coal accounted for 4 quadrillion Btu of U.S. energy in 2007.
   (3) Nuclear power is a major contributor of U.S. energy.
   (4) Petroleum, natural gas, and coal supply the largest portion of all U.S. energy.
   (5) U.S. reliance on petroleum is decreasing each year.

UNIT 3

3 Apply the Skill

Directions: Choose the one best answer to each question.

Questions 2 and 3 refer to the following pictograph.

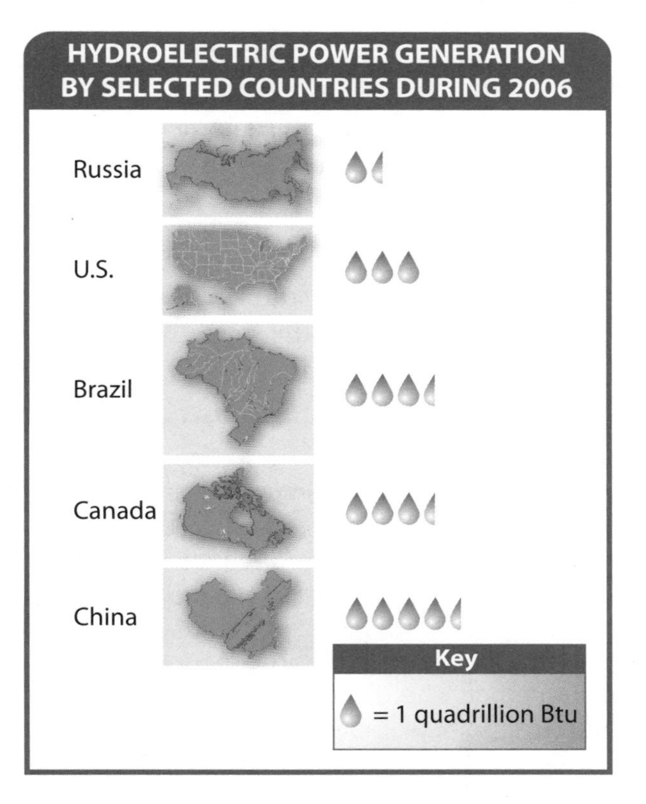

2. Which of the following conclusions can be drawn from information presented in the pictograph?

   (1) Russia produced more hydroelectric power than the United States did in 2006.
   (2) China produced all of its electricity from hydroelectric power plants in 2006.
   (3) In 2006, Brazil and the United States produced the same amount of hydroelectric power.
   (4) China produced more than twice as much hydroelectric power as Russia in 2006.
   (5) Canada produced more hydroelectric power in 2006 than Russia and Brazil combined.

3. Which of the following is the best estimate of Brazil's hydroelectric power generation in 2006?

   (1) 3 trillion Btu
   (2) 3.5 trillion Btu
   (3) 3.5 quadrillion Btu
   (4) 3 quadrillion Btu
   (5) 4 quadrillion Btu

Questions 4 and 5 refer to the following paragraph and pictograph.

Electric power represents one portion of the total energy picture. The amount of electricity consumed by a homeowner is measured in kilowatt-hours. A typical homeowner observes a change in his or her kilowatt-hour use from one season to the next. In the southwestern United States, summertime brings higher electric bills as people run air conditioning units more frequently than they do in winter months.

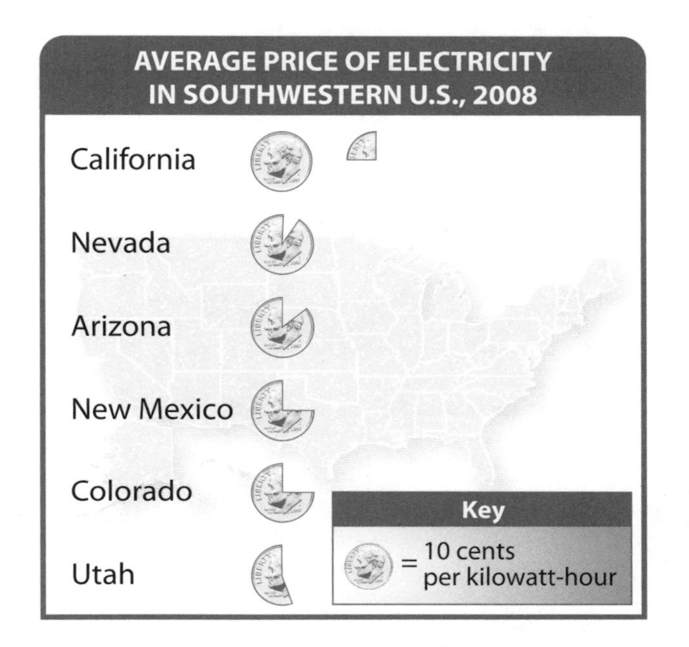

4. Based on the information, what can you conclude about the average price of electricity in the southwestern United States?

   The average price of electricity

   (1) is higher in Colorado than in Nevada
   (2) is about the same in Colorado and New Mexico
   (3) is about 20 cents per kilowatt-hour in Arizona
   (4) costs more in Utah than in California
   (5) is less than 5 cents per kilowatt-hour in Utah

5. Based on the pictograph, what is the average price of electricity per kilowatt-hour in Utah?

   (1) about 5 cents
   (2) about 6 cents
   (3) about 10 cents
   (4) about 12 cents
   (5) about 15 cents

UNIT 3

LESSON 10

Relate Text and Figures

1 Learn the Skill

As you know, illustrations, graphs, and tables present information in a visual way. Text can help you understand information in these graphics. In this way, text and graphics support one another. Knowing how to **relate text and figures** will allow you to fully understand the information being presented.

2 Practice the Skill

By mastering the skill of relating text and figures, you will improve your study and test-taking skills, especially as they relate to the GED Science Test. Study the paragraph, diagram, and strategies below. Then answer the question that follows.

A When you see labels in a diagram, also look for those words in the text. The text may offer additional explanations that can help clarify the labels.

B When text and a graphic occur together, both are usually required to answer the question. Here, read the text carefully. Notice that the text only discusses transverse waves and the diagram shows a transverse wave. Other types of waves are not discussed.

UNIT 3

When waves travel through substances, they cause the particles in substances to vibrate. In transverse waves, **B** the particles vibrate perpendicular to the direction in which the wave is moving. The diagram shows how the particles move away from their resting position when a transverse wave passes through a substance. Crests **A** and troughs are the points where particles are farthest from their resting point. The distance from any one point to the next identical point is known as the wavelength.

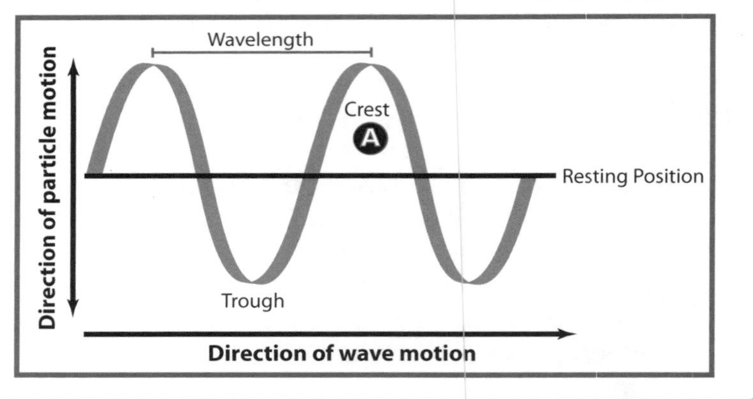

MAKING ASSUMPTIONS

Whenever a graphic accompanies text, you can assume the two pieces of information are related. You also should assume that you need both pieces of information to answer the questions.

1. Based on the information in the paragraph and diagram, which statement is true?

   (1) The wavelength can only be measured from crest to crest.
   (2) The larger the crest of a wave, the farther the particles move from their resting point.
   (3) Transverse waves cause particles to move in the same direction as the wave motion.
   (4) The wavelength measures how far the particles move from their resting point.
   (5) All waves cause particles to vibrate perpendicular to the direction in which the wave moves.

③ Apply the Skill

Directions: Choose the one best answer to each question.

Questions 2 and 3 refer to the following paragraphs and diagram.

Sound waves are mechanical waves because they can only move through substances, such as air. They are longitudinal waves because they cause substances to vibrate back and forth parallel to the direction in which the wave is moving. When a sound wave travels through air, it causes the particles in air to move back and forth.

The amplitude of a mechanical wave is the point at which the particles move the farthest from their resting point. Loud sounds are caused by waves that have large amplitudes. Pitch describes how high or low a sound is. The wavelength of a sound wave determines the pitch.

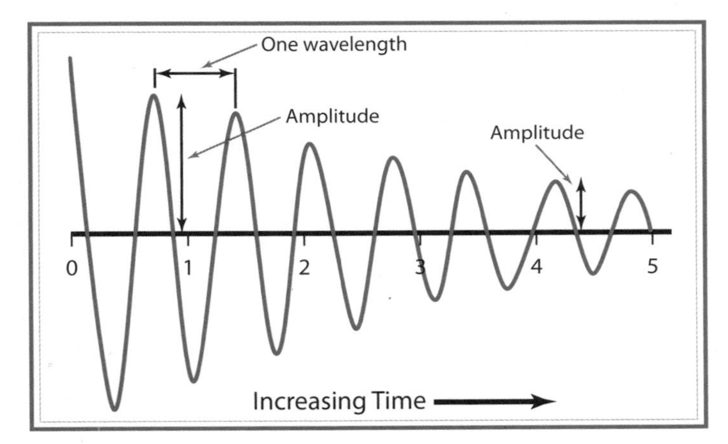

2. Based on the information, which of the following statements is the most accurate?

   A sound wave that has a small amplitude will produce a sound that

   (1) has a high pitch
   (2) has a low pitch
   (3) has a large wavelength
   (4) is loud
   (5) is soft

3. What is true of the sound that is represented by the diagram?

   (1) Its pitch increase over time.
   (2) Its loudness is constant.
   (3) It becomes softer over time.
   (4) It can travel in the absence of air.
   (5) It can travel only through air.

Questions 4 and 5 refer to the following paragraph and table.

The volume of sound is related to wave amplitude. The larger the amplitude, the greater the volume. Volume is measured in decibels (dB). Hearing damage begins at about the 85 dB level. The table lists the decibel levels of some sounds.

VOLUME OF SOUND	
SOUND	**DECIBEL LEVEL (dB)**
Conversation	50–65
Vacuum cleaner	70
Lawn mower	85–90
Jackhammer	110
Jet engine	140

4. Based on the information in the paragraph and table, which sounds can cause hearing damage?

   (1) a conversation, a vacuum cleaner, and a lawn mower
   (2) a vacuum cleaner and a lawn mower
   (3) a lawn mower, a jackhammer, and a jet engine
   (4) a vacuum cleaner, a jackhammer, and a jet engine
   (5) a conversation, a vacuum cleaner, a lawn mower, a jackhammer, and a jet engine

5. Based on the information, sound waves from which source have the smallest amplitude?

   (1) conversation
   (2) vacuum cleaner
   (3) lawn mower
   (4) jackhammer
   (5) jet engine

UNIT 3

Analyze Results

1 Learn the Skill

The process known as the scientific method usually begins with a question. Using what they already know, scientists attempt to answer the question with a hypothesis, or an educated guess. Scientists then try and prove their hypothesis by conducting tests and gathering data. After the testing is complete, the scientists subject their data to analysis. **Analyzing results** helps scientists determine whether their initial hypothesis is correct.

2 Practice the Skill

By mastering the skill of analyzing results, you will improve your study and test-taking skills, especially as they relate to the GED Science Test. Examine the graphic and paragraph below. Then answer the question that follows.

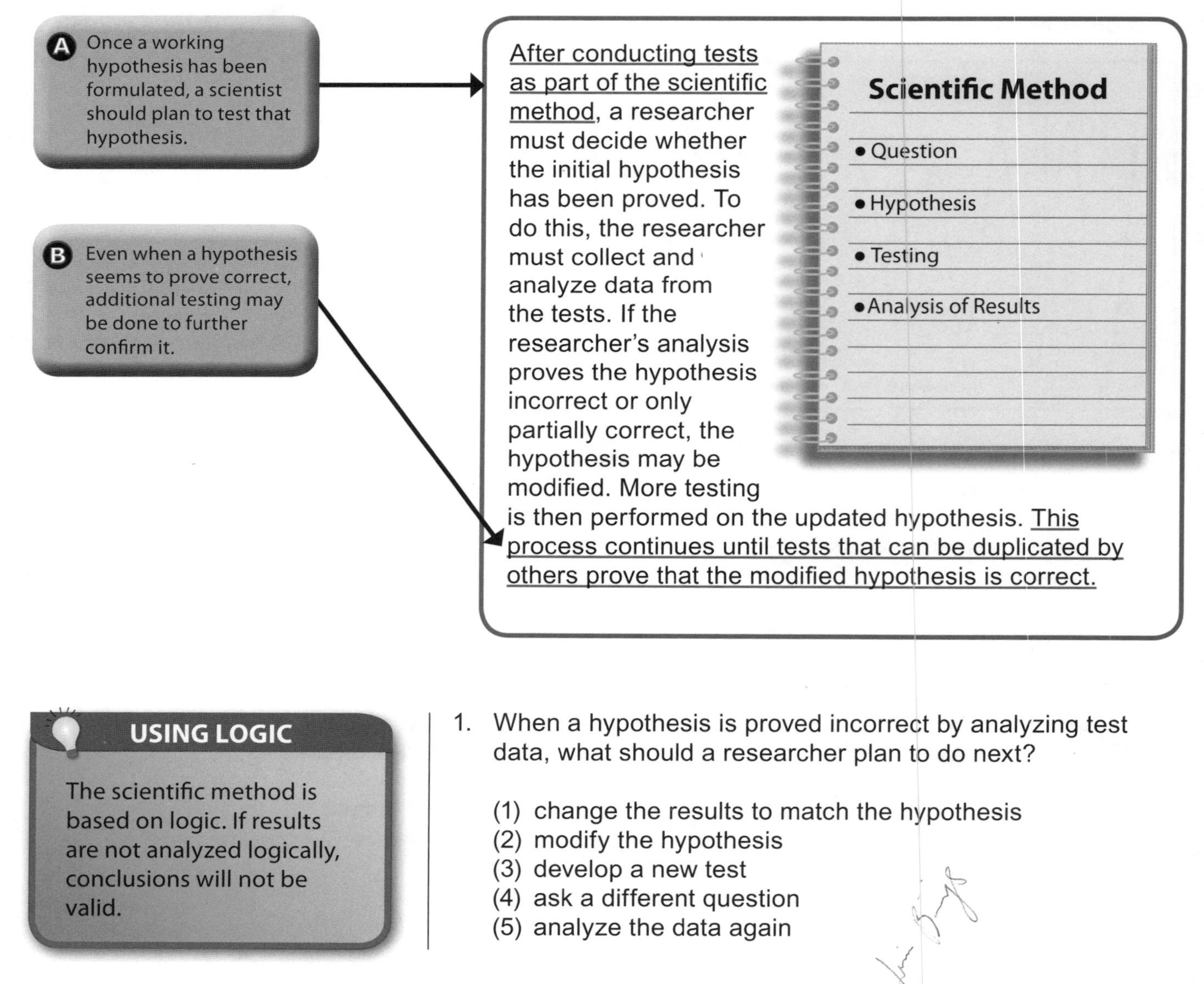

After conducting tests as part of the scientific method, a researcher must decide whether the initial hypothesis has been proved. To do this, the researcher must collect and analyze data from the tests. If the researcher's analysis proves the hypothesis incorrect or only partially correct, the hypothesis may be modified. More testing is then performed on the updated hypothesis. This process continues until tests that can be duplicated by others prove that the modified hypothesis is correct.

USING LOGIC

The scientific method is based on logic. If results are not analyzed logically, conclusions will not be valid.

1. When a hypothesis is proved incorrect by analyzing test data, what should a researcher plan to do next?

   (1) change the results to match the hypothesis
   (2) modify the hypothesis
   (3) develop a new test
   (4) ask a different question
   (5) analyze the data again

③ Apply the Skill

Directions: Choose the one best answer to each question.

Questions 2 through 6 refer to the following paragraph and graphic.

A scientist wants to know whether certain common liquids have the acidic potential to kill some varieties of aquatic life. The scientist forms a hypothesis stating that five common liquids are acidic enough to be toxic to fish and young frogs. While conducting background research, the scientist learns that U.S. government researchers had previously studied the effects of various pH levels on some common forms of aquatic life.

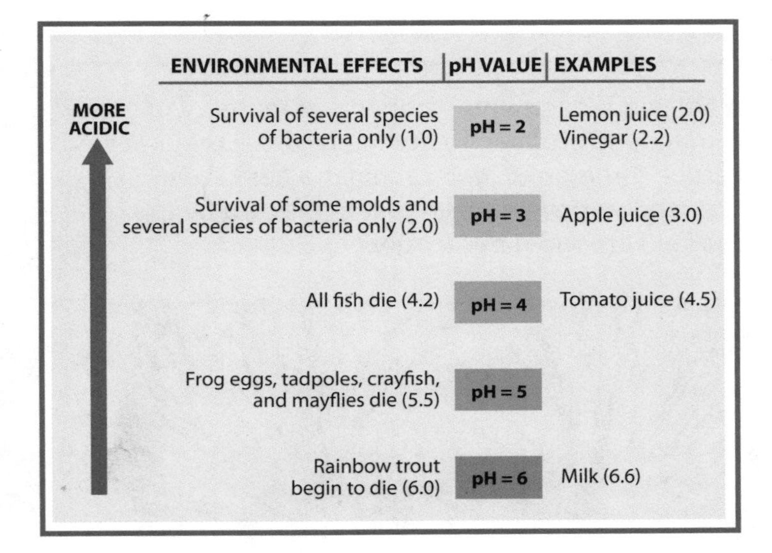

2. Which of the following conclusions is a logical analysis of the test data?

   (1) None of the tested liquids threatens aquatic life.
   (2) All of the tested liquids except lemon juice threaten aquatic life.
   (3) All of the tested liquids threaten aquatic life.
   (4) All of the tested liquids except milk threaten aquatic life.
   (5) Only tomato juice threatens aquatic life.

3. To which of the following further studies would this data be most relevant?

   (1) the effects of acid rain on food chains
   (2) the effects of high acidity on the human digestive system
   (3) the possible role of tomato juice as a nutrient source for aquatic organisms
   (4) an analysis of food preferences of tadpoles
   (5) an analysis of predator-prey relationships in pond ecosystems

4. Does analysis of the test data support the hypothesis?

   (1) Yes, it completely supports the hypothesis.
   (2) No, it completely refutes the hypothesis, and a new hypothesis must be formed.
   (3) It demonstrates a flaw in the hypothesis, so the hypothesis must be modified.
   (4) The data are not sufficient to confirm or refute the hypothesis.
   (5) The data do not relate to the background research.

5. Based on the analysis of the test results, which of the following is the next step in the scientific method?

   The researcher should now

   (1) modify her hypothesis and repeat her tests
   (2) retest to confirm her hypothesis, since it has been supported
   (3) develop an entirely new hypothesis and repeat the same tests
   (4) develop an entirely new hypothesis and conduct entirely new tests
   (5) discontinue her research

6. Based on the data, which organisms are least tolerant of high acidity?

   (1) tadpoles
   (2) crayfish
   (3) mayflies
   (4) trout
   (5) frog eggs

Apply Concepts

1 Learn the Skill

A **concept** is a model based on general knowledge of a subject. Concepts can be useful in determining answers to new questions or in finding solutions to new problems. Researchers often interpret new information by relying on concepts with which they are already familiar. Knowing how to **apply concepts** is valuable to research and discovery.

2 Practice the Skill

By mastering the skill of applying concepts, you will improve your study and test-taking skills, especially as they relate to the GED Science Test. Examine the paragraph and diagram below. Then answer the question that follows.

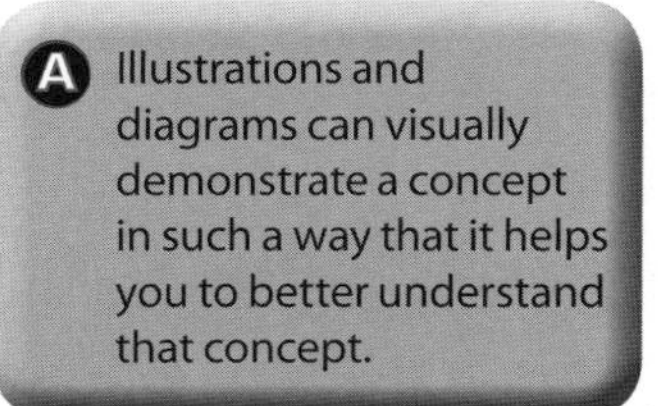
A Illustrations and diagrams can visually demonstrate a concept in such a way that it helps you to better understand that concept.

B Text in a question often gives background information about a concept. It may also give some details about the concept.

Magnetism is a fundamental property of matter. Objects that are magnetic produce magnetic fields. A magnetic field exerts a force on objects within the field. The illustration shows how iron filings are affected by the magnetic field around a bar magnet.

A

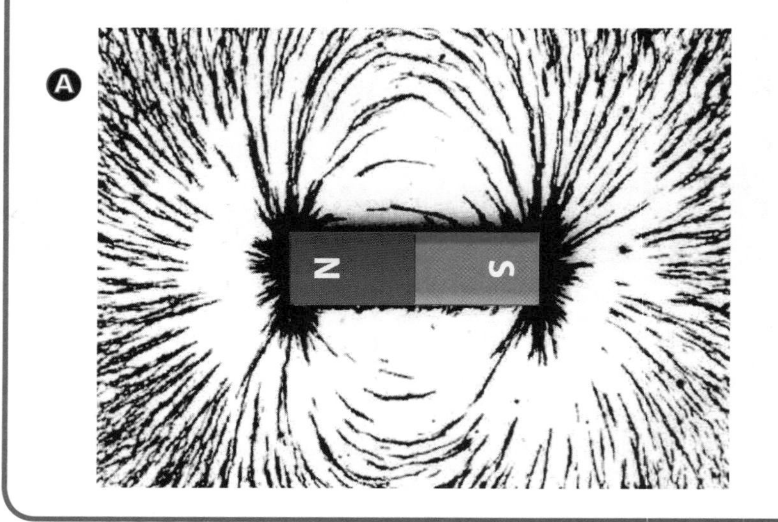

MAKING ASSUMPTIONS

You can assume that processes and forces described in one situation apply to other similar situations. Making this assumption allows you to apply known concepts to new situations.

1. Based on the information, where is the bar magnet's **B** magnetic field most likely strongest?

   (1) in the center of the magnet
   (2) several inches away from the magnet
   (3) along a circle through the center of the magnet
   (4) at the magnet's north and south poles
   (5) directly above the magnet

③ Apply the Skill

Directions: Choose the one best answer to each question.

Questions 2 through 5 refer to the following paragraph and illustrations.

Magnets are surrounded by invisible lines of force. These lines result from the property of bipolarity. Bipolarity means that a magnet has two poles where the lines of force are most concentrated. These poles are designated "north" and "south." Force lines moving from north to south create a magnet's field. When two magnets are brought close to each other, their magnetic fields interact. The directions of their force lines result in predictable patterns of attraction and repulsion.

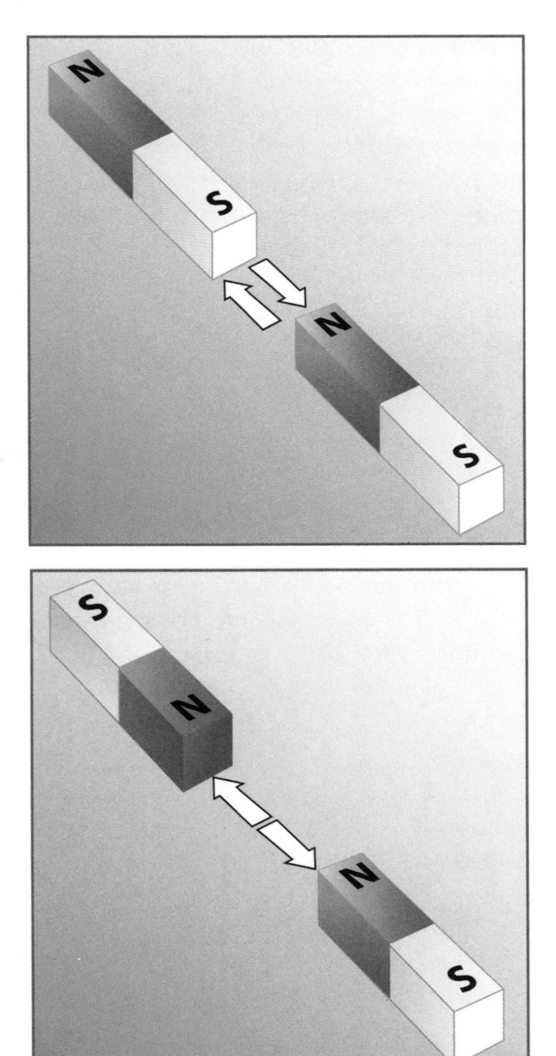

2. Based on your own knowledge and the information, which of the following statements is true?

   The illustration demonstrates that

   (1) opposite magnetic poles repel each other, while like poles attract each other
   (2) opposite magnetic poles attract each other, while like poles repel each other
   (3) all magnets always attract each other
   (4) all magnets always repel each other
   (5) magnetic forces surround all matter

3. Which of the following concepts is supported by the text and illustrations?

   (1) Magnetic fields are perfectly circular.
   (2) Magnetic forces are variable.
   (3) Magnetic forces have a great deal of strength.
   (4) Magnetic forces do not affect solid objects.
   (5) Magnetic lines of force are directional.

4. Based on your own knowledge and the information, which of the following statements is true?

   The reaction of two magnets placed side by side depends on

   (1) the metal of which each magnet is made
   (2) the temperature of the surrounding air
   (3) which of the magnets is placed on the left side
   (4) how the magnets' poles are aligned
   (5) which of the magnets has bipolarity

5. Which of the following would be the best way to see the range of a magnet's field?

   (1) submerge it in water
   (2) sprinkle sand on and around it
   (3) sprinkle iron shavings on and around it
   (4) place a piece of metal near it
   (5) place another magnet near it

UNIT 3

Unit 3 Review

The Unit Review is structured to resemble the GED Science Test. Be sure to read each question and all possible answers very carefully before choosing your answer.

To record your answers, fill in the numbered circle that corresponds to the answer you select for each question in the Unit Review.

Do not rest your pencil on the answer area while considering your answer. Make no stray or unnecessary marks. If you change an answer, erase your first mark completely.

Mark only one answer space for each question; multiple answers will be scored as incorrect.

Sample Question

In which of the following examples would a pictograph provide the best visual representation of information?

(1) when showing the concentration of water molecules in different states
(2) when showing ionization reactions
(3) when showing energy consumption of people in the United States
(4) when showing change over time in energy consumption in the United States
(5) when showing the relative positions of atoms in a molecule

Directions: Choose the one best answer to each question.

Questions 1 and 2 refer to the following paragraph and table.

The function of a machine is to transfer or change a force. Any machine is made up of one or more of the six simple machines. Simple machines can be characterized by the mechanical advantage that they provide. Mechanical advantage (MA) is a ratio of the size of the output force to the size of the input force.

$$MA = \frac{\text{output force}}{\text{input force}}$$

MECHANICAL ADVANTAGE

MACHINE	INPUT FORCE (N)	OUTPUT FORCE (N)
A	300	900
B	15	150
C	5	10
D	800	1000
E	10	40

1. Which machine provides the greatest mechanical advantage?

(1) A
(2) B
(3) C
(4) D
(5) E

①②③④⑤

2. The output force of a particular simple machine is 100 N. The machine provides a mechanical advantage of 4. What is the input force?

(1) 4 N
(2) 25 N
(3) 100 N
(4) 200 N
(5) 400 N

①②③④⑤

Questions 3 and 4 refer to the following paragraphs.

The everyday meanings of the words *hypothesis* and *theory* are different from their scientific meanings. When reading and evaluating scientific information, it is important to understand what scientists mean when they use these words.

In science, a hypothesis is an educated guess that is based on observations and knowledge. It is a possible answer to a scientific question. A hypothesis can be proved false, but it can never be proved true, because there is no way to test every possible situation to see whether the hypothesis holds true in all of them. A hypothesis generally applies only to a very specific situation.

A scientific theory is a well-supported explanation for many different observations. For a scientific theory to be accepted, all available observations and evidence must support it. A valid scientific theory, like a hypothesis, can be proved false if enough observations contradict it. An important characteristic of both hypotheses and theories is that they *must* be disprovable. In other words, a statement is not a valid scientific hypothesis or theory if there is no possible way to prove that it is not true.

3. Consider the following statement:

   "The force of gravity can be explained by the presence of invisible rays that objects create. The rays interact with one another, producing gravity. Because the rays are neither matter nor energy, there can be no way to detect them."

   Which of the following statements best explains why the statement above is not a valid scientific hypothesis?

   (1) It does not explain observable events.
   (2) There is no way to test it in all situations.
   (3) There is no test that could prove it is false.
   (4) No scientists accept it as true.
   (5) It relies on things that cannot be detected with the senses.

   ①②③④⑤

4. The modern atomic theory was developed in the late 1800s and early 1900s. It replaced earlier atomic theories. What was the most likely reason scientists accepted the modern atomic theory and rejected older atomic theories?

   (1) The modern atomic theory explains new observations that older theories could not.
   (2) The scientists who proposed the modern atomic theory were better respected than those who supported older theories.
   (3) Scientists knew that newer theories are always more accurate than older theories.
   (4) Most of the general public thought that the atomic theory was correct.
   (5) The modern atomic theory was published in more scientific journals than older theories had been.

   ①②③④⑤

Question 5 refers to the following paragraph and diagram.

According to Newton's second law of motion, the motion of an object can change only if unbalanced forces are acting on it.

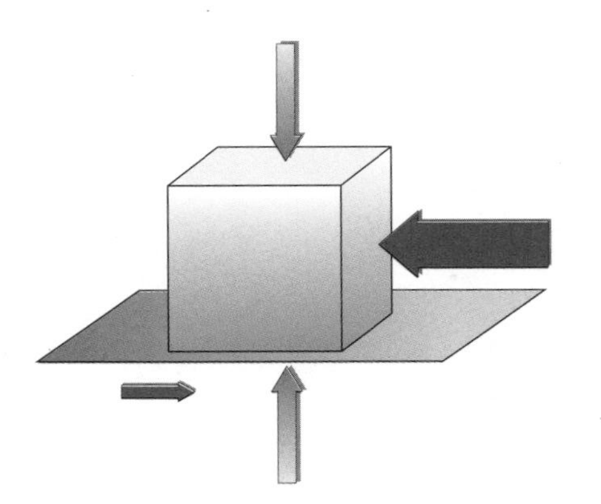

5. Based on the information, which statement describes the motion of the box when the forces represented in the diagram act on it?

   The box will

   (1) remain in place
   (2) move far to the left
   (3) move slightly left
   (4) move far to the right
   (5) move slightly right

   ①②③④⑤

UNIT 3

Questions 6 and 7 refer to the following text and graph.

A child's toy is rolling from the top of a steep ramp (point 25) to the bottom (point 0). The graph represents the potential and kinetic energy at various points on the ramp.

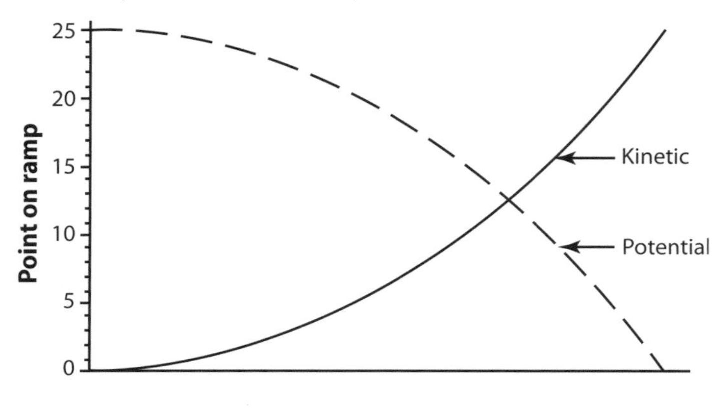

6. At which point(s) is the potential energy of the toy at its greatest?

   Potential energy of the toy is greatest at point(s)

   (1) 0
   (2) 5
   (3) 10
   (4) 25
   (5) 0 and 25

①②③④⑤

7. Which of the following conclusions is best supported by data in the graph?

   (1) As potential energy decreases, kinetic energy increases.
   (2) An object can have potential energy or kinetic energy, but not both.
   (3) An increase in potential energy causes an increase in kinetic energy.
   (4) Kinetic energy is lowest when an object is in motion.
   (5) As potential energy decreases, kinetic energy decreases.

①②③④⑤

Question 8 refers to the following paragraphs and illustrations.

Although many people don't realize it, electricity and magnetism are closely related. A magnet spinning inside a coil of wire produces an electric current in the wire. Similarly, running electric current through a wire produces a magnetic field around the wire. The direction in which the current flows determines where the north and south poles of the magnetic field lie.

This relationship is very important to many aspects of modern life. For example, most of the electricity we use is generated by moving magnets. Electric power plants use different forms of energy—such as moving water or steam—to make large magnets turn. The turning magnets produce electric current, which is transmitted by wires to homes and businesses.

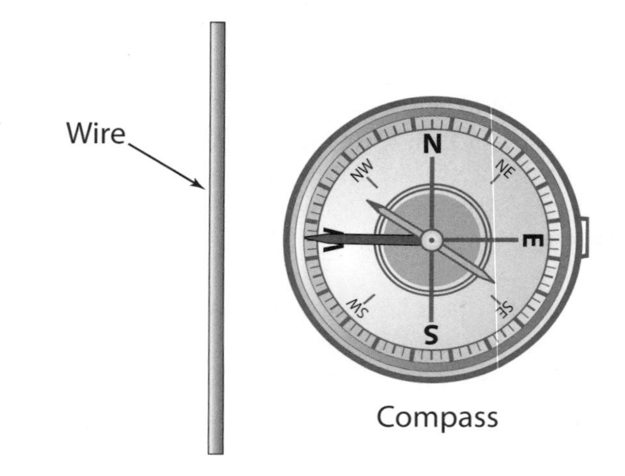

8. There is an electric current in the wire shown in the illustration. If the direction of that electric current were reversed, what would most likely happen to the needle on the compass?

   The compass needle would

   (1) spin around quickly
   (2) point to the left
   (3) point to the right
   (4) point downward
   (5) point upward

①②③④⑤

Questions 9 and 10 refer to the following paragraphs and diagrams.

Most waves can be classified into one of two types: transverse waves or longitudinal waves. Transverse waves are the waves that come to mind when many people think of waves. When a transverse wave passes through matter, it moves the matter up and down or side to side. The matter moves in a direction perpendicular to the direction that the wave is traveling. Part A of the diagram below shows the parts of a transverse wave.

When a longitudinal wave passes through matter, it causes the matter to expand and contract. The matter moves in a direction parallel to the direction that the wave is traveling. Part B of the diagram shows a longitudinal wave.

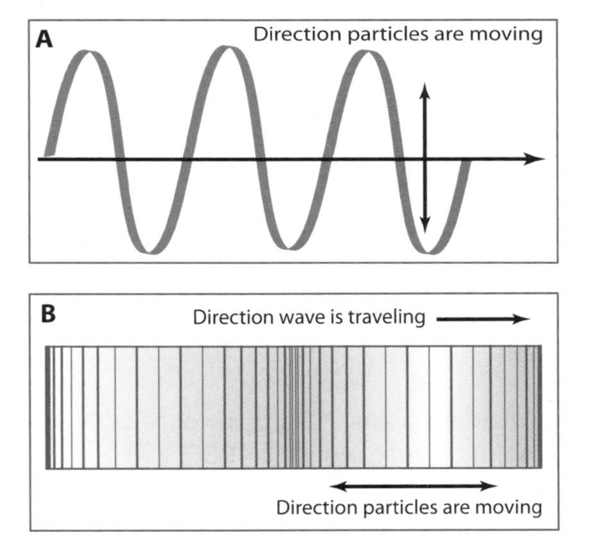

9. Based on the information, in what direction is the transverse wave in Part A of the diagram most likely moving?

   The wave is moving

   (1) outward from a central point
   (2) upward
   (3) from the edge of the wave to its center
   (4) along a curved path
   (5) from left to right

   ①②③④⑤

10. Sound waves are longitudinal waves. If a sound wave passes through an object from top to bottom, what type of motion do the particles in the object probably experience?

    The particles in the object move

    (1) up and down
    (2) left and right
    (3) front to back
    (4) along a circular path
    (5) randomly

    ①②③④⑤

Question 11 refers to the following paragraph, table, and models.

Organic chemistry is the study of compounds that contain chains of carbon atoms. Prefixes are commonly used in the names of organic compounds to indicate the number of carbon atoms in the main carbon chain. A group of organic compounds called alcohols all contain an –OH group attached somewhere along the main (longest) carbon chain. The models below represent molecules of two alcohols.

PREFIXES AND THE NUMBERS THEY REPRESENT

PREFIX	NUMBER	PREFIX	NUMBER
Meth-	1	*But-*	4
Eth-	2	*Pent-*	5
Prop-	3	*Hex-*	6

Molecule A　　Molecule B

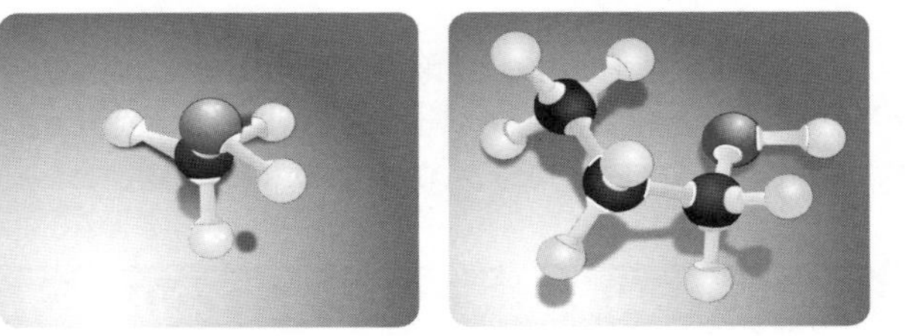

11. Based on the information, what is the name of Molecule B?

    (1) ethanol
    (2) propanol
    (3) butanol
    (4) pentanol
    (5) hexanol

    ①②③④⑤

Question 12 refers to the following pictograph.

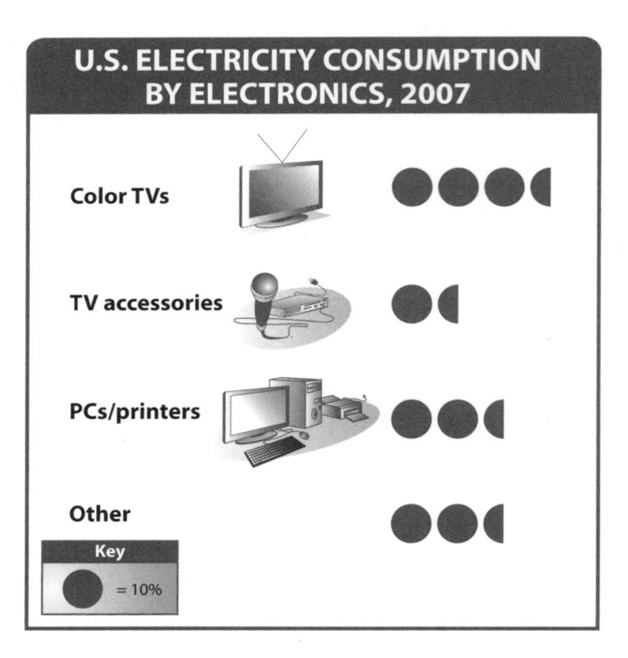

12. About what percentage of electricity is used per year in the average U.S. household for TVs and TV accessories?

(1) 10%
(2) 15%
(3) 25%
(4) 30%
(5) 50%

①②③④⑤

Questions 13 and 14 refer to the following paragraph and chemical equations.

The process in which atoms and molecules are arranged into new combinations is called a chemical reaction. The table below identifies several different types of chemical reactions.

REACTION TYPE	MODEL EQUATION
Synthesis	$A + B \rightarrow AB$
Decomposition	$AB \rightarrow A + B$
Single displacement	$AX + B \rightarrow BX + A$
Double displacement	$AX + BY \rightarrow AY + BX$
Neutralization	$AOH + HB \rightarrow AB + H_2O$

13. Which of the following reactions would be classified as a single-displacement reaction? (Note: The equations given below are not balanced.)

(1) $CuCl_2 + Al \rightarrow AlCl_3 + Cu$
(2) $Na + Cl_2 \rightarrow NaCl$
(3) $HCl + NaOH \rightarrow NaCl + H_2O$
(4) $S_8 + O_2 \rightarrow SO_2$
(5) $NaCl + H_2SO_4 \rightarrow Na_2SO_4 + HCl$

①②③④⑤

14. Which kind of reaction is represented by the chemical equation below?

$$Pb(NO_3)_2 + KI \rightarrow KNO_3 + PbI_2$$

(1) synthesis
(2) decomposition
(3) single displacement
(4) double displacement
(5) neutralization

①②③④⑤

Question 15 refers to the following information and graph.

The melting point of pure water is 0°C. The boiling point of pure water is 100°C.

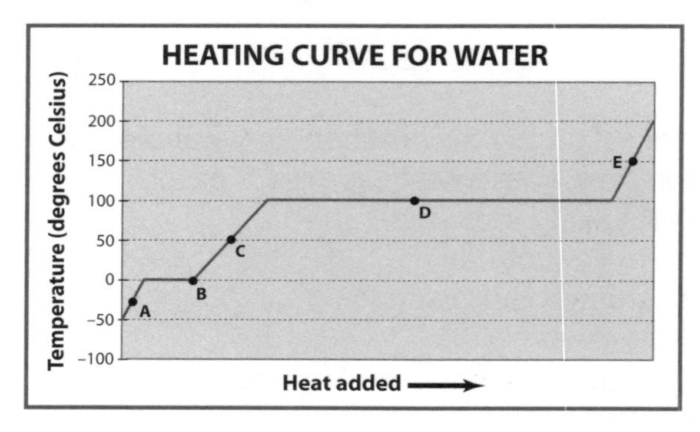

15. Based on the information, which state(s) of water would you expect to have in a sample that is at Point D on the heating curve?

(1) solid
(2) liquid
(3) gas
(4) solid and liquid
(5) liquid and gas

①②③④⑤

Questions 16 and 17 refer to the following paragraph and chart.

The pH of a solution describes the concentration of hydronium (H_3O^+) ions. The greater the concentration of hydronium ions, the more acidic a solution becomes. A pH value also indicates the concentration of hydroxide (OH^-) ions. The greater the concentration of hydroxide ions, the more basic a solution becomes.

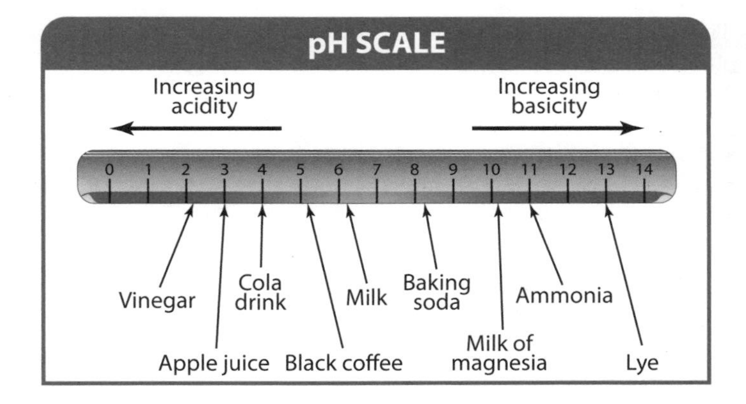

16. Which of the following substances is the most acidic?

(1) milk of magnesia
(2) baking soda
(3) cola drink
(4) apple juice
(5) vinegar

①②③④⑤

17. Which of the following solutions has the lowest concentration of hydronium ions?

(1) black coffee
(2) ammonia
(3) cola drink
(4) milk
(5) vinegar

①②③④⑤

Questions 18 and 19 refer to the following paragraph and table.

Acids and bases are two categories of chemical compounds. Litmus paper is one indicator used to determine whether a compound is an acid or base. Acids turn blue litmus paper red. Bases turn red litmus paper blue. Combining an acid and a base may produce water and a salt. Such a reaction is called a neutralization reaction.

ACID AND BASE COMBINATIONS

COMPOUND	FORMULA	RESULT OF LITMUS TEST
Nitric acid	HNO_3	Blue litmus turns red
Sodium hydroxide	NaOH	Red litmus turns blue
Barium hydroxide	$Ba(OH)_2$	Red litmus turns blue
Sulfuric acid	H_2SO_4	Blue litmus turns red
Potassium hydroxide	KOH	Red litmus turns blue
Perchloric acid	$HClO_4$	Blue litmus turns red

18. Which of the following pairs of compounds could be involved in a neutralization reaction?

(1) HNO_3 and $HClO_4$
(2) $HClO_4$ and HNO_3
(3) NaOH and H_2SO_4
(4) NaOH and KOH
(5) $Ba(OH)_2$ and KOH

①②③④⑤

19. A scientist tests a solution with litmus paper. The solution turns red litmus paper blue. Which of the following substances is most likely present in the solution?

(1) magnesium hydroxide, $Mg(OH)_2$
(2) acetic acid, HCH_3COO
(3) sodium bromide, NaBr
(4) hydrochloric acid, HCl
(5) iron oxide, Fe_2O_3

①②③④⑤

UNIT 3

Answer Key

UNIT 1 LIFE SCIENCE

LESSON 1, *pp. 2–3*

1. Analysis: (3), Earthworms are annelids, and annelids have segments that are all about the same size.
2. Application: (4), The table gives information about the defining characteristics of the five main types of chordates.
3. Application: (5), All chordates have limbs.
4. Evaluation: (5), Both birds and reptiles have lungs and lay eggs with shells.
5. Comprehension: (4), Four of the five chordate groups use lungs at some point in their life cycles.
6. Application: (3), Information on habitat is the only information not given in the table.

LESSON 2, *pp. 4–5*

1. Comprehension: (4), Eighty-five percent of the bear's diet comes from fruit and plants.
2. Analysis: (3), The panda population was smaller in 2007 than in 1975.
3. Analysis: (4), The gray wolf's population has increased since 1974, when the Endangered Species Act was passed.
4. Comprehension: (1), The population of gray wolves was lower in 1973 than in any other year shown on the graph.

LESSON 3, *pp. 6–7*

1. Analysis: (4), Arrows in a food chain point from prey to predator.
2. Comprehension: (5), Caterpillars appear in the overlap between the two circles.
3. Evaluation: (1), The diagram shows foods eaten by two desert animals.
4. Comprehension: (2), Other mice appear in the circle describing the diet of grasshopper mice.
5. Analysis: (3), Arrows in a food web point from prey to predator.
6. Analysis: (1), The food web shows that many animals in a grassland eat grass.

LESSON 4, *pp. 8–9*

1. Comprehension: (4), The diagram indicates that ribosomes are involved in protein production.
2. Comprehension: (5), The diagram shows the lysosome floating in the cytoplasm.
3. Application: (1), The nucleus is responsible for cellular reproduction.
4. Application: (4), The electron microscope, invented in the 1950s, allowed scientists to see the internal structures of cells.
5. Analysis: (3), The diagram shows that the axon is involved in transmitting information.
6. Evaluation: (4), Dendrites help carry electrical signals through the body.

LESSON 5, *pp. 10–11*

1. Application: (2), Using bacteria to produce medications is an example of a beneficial use of bacteria.
2. Analysis: (3), The table groups bacteria by shape.
3. Evaluation: (5), This statement describes two different effects of different types of bacteria.
4. Analysis: (3), The passage states that bacteria help convert nitrogen to a form plants can use.
5. Evaluation: (1), Both the diagram and the passage describe different ways that nitrogen changes form in the environment.

LESSON 6, *pp. 12–13*

1. Analysis: (1), The passage states that there are many different kinds of plants, and that most organisms rely on plants to survive.
2. Analysis: (1), The diagram describes the names and roles of the main parts of a vascular plant.
3. Analysis: (5), The passage describes how vascular and nonvascular plants move materials through their bodies.
4. Application: (2), The passage and diagram focus mainly on the movement of substances through plants.
5. Application: (2), The passage describes how cactuses differ from many other plants.

LESSON 7, *pp. 14–15*

1. Application: (2), The passage states that fertilizers contribute to algal blooms.
2. Evaluation: (4), The diagram shows that riparian areas are located near water, and the passage implies that riparian areas are located near water.
3. Analysis: (1), The passage describes the benefits of wetlands and riparian areas.
4. Application: (2), The passage describes the monitoring and control of common pollutants.

LESSON 8, *pp. 16–17*

1. Application: (2), Deserts, grasslands, and temperate deciduous forests make up most of the United States; deserts receive an average of 25 cm of rain each year, grasslands receive about 65 cm of rain each year, and temperate deciduous forests receive about 125 cm of rain each year.
2. Analysis: (4), Boreal forest and tundra are the coldest of the biomes.
3. Analysis: (3), In order to survive, plants and animals adapt to the biome in which they live.
4. Application: (3), Animals in boreal forests and tundras must have adaptations for cold weather.
5. Evaluation: (1), Boston and Paris are located in the same type of biome.

LESSON 9, *pp. 18–19*

1. Analysis: (4), The table shows examples of organisms in two different kingdoms.
2. Evaluation: (1), Classes are divided into orders, and orders are divided into families; therefore, each class contains at least one family, so there are more families than classes.
3. Application: (3), Plants can make their own food.

UNIT 1 *(continued)*

LESSON 10, *pp. 20–21*

1. Comprehension: (5), All mammals have fur or hair.
2. Analysis: (3), The table shows that all three classes of mammals make milk for their young.
3. Analysis: (5), The table shows that monotremes are the only mammals that lay eggs.
4. Analysis: (2), The diagram states that lions hunt, but kangaroos do not.
5. Evaluation: (3), Both placental mammals and marsupials give birth to live young.

LESSON 11, *pp. 22–23*

1. Comprehension: (4), After the stomach, food passes into the small intestine.
2. Comprehension: (3), The sight and smell of food stimulate saliva production.
3. Analysis: (4), Food must enter the mouth and be pushed to the back of the mouth before it can be chewed.
4. Evaluation: (1), The flow chart shows that bile and pancreatic juices enter the small intestine.
5. Evaluation: (3), The diagram shows that wastes leave the small intestine, so digestion is mostly complete after food moves through the small intestine.

LESSON 12, *pp. 24–25*

1. Application: (3), Conservation laws were still being passed after the 1950s.
2. Analysis: (5), All three species became extinct within a ten-year period.
3. Evaluation: (4), DDT caused reproductive problems in bald eagles between the 1940s and the 1960s.
4. Evaluation: (5), Government mandates in the 1960s and 1970s led to the recovery of bald eagles in the 21st century.

LESSON 13, *pp. 26–27*

1. Analysis: (3), The diagram shows that fluid and blood cells move into the affected area, which produces swelling.
2. Comprehension: (2), Exposure to cowpox produces antibodies to cowpox virus; these antibodies also provide some protection against smallpox.
3. Comprehension: (5), Bacterial infection causes the immune system to produce proteins to destroy the bacteria.
4. Application: (5), Pasteur's germ theory of disease revolutionized medicine.

LESSON 14, *pp. 28–29*

1. Evaluation: (4), Darwin infers that smaller eyes would be less prone to inflammation and therefore would be beneficial to burrowing animals.
2. Analysis: (4), Darwin assumed that the finch species evolved from a single species.
3. Comprehension: (3), Darwin used his observations to learn how finches may have come to exist.
4. Application: (4), Beak size is related to the type of food that each bird eats; since Bird A and Bird B have different-sized beaks, they probably eat different foods.
5. Application: (1), The land on which the mussel beds were found must have been located below the high-tide level prior to the earthquake.

LESSON 15, *pp. 30–31*

1. Analysis: (2), The white-flower trait is not present in the first generation, but it reappears in the second generation.
2. Analysis: (4), Only plants with two copies of the white-flower trait produce white flowers.
3. Application: (2), Although both parent flowers had one copy of the purple-flower trait and one copy of the white-flower trait, they produced flowers that were both purple and white in color.
4. Evaluation: (2), A pea plant has white flowers only if both its parents pass on a copy of the white-flower trait.

LESSON 16, *pp. 32–33*

1. Analysis: (2), The paragraph states that most producers use sunlight to make food.
2. Evaluation: (2), A community is made up of different populations, and each population is made up of a single species.
3. Application: (2), Communities in different ecosystems contain different populations.
4. Comprehension: (1), Only a few types of organisms can live in very hot or very cold environments.
5. Comprehension: (3), The paragraph states that most organisms survive best under a narrow range of conditions.

LESSON 17, *pp. 34–35*

1. Application: (4), In a mutualistic relationship, both species benefit.
2. Analysis: (3), The orchids must benefit for the relationship to be commensal.
3. Application: (2), In a mutualistic relationship, both species benefit.
4. Application: (2), The graph shows how changes in one population affect the size of the other population.
5. Application: (1), If this graph is similar to graphs of other predator-prey relationships, this graph could be used to learn more about predator-prey relationships.

UNIT 1 REVIEW, *pp. 36–41*

1. Analysis: (3), At point X, the slope of the curve becomes zero, so the population is no longer growing.
2. Application: (1), A limited food supply would restrict the population of a species to a certain level.
3. Comprehension: (1), The paragraph describes two types of plant adaptations.
4. Analysis: (3), Denitrifying bacteria convert nitrate into gaseous nitrogen.
5. Analysis: (3), Both parents had one *p* trait.
6. Evaluation: (4), Three-quarters of the offspring will have at least one purple-flower trait.
7. Application: (2), Kelp make their own food.
8. Analysis: (5), No arrow points from the kelp to the sea star.

UNIT 1 *(continued)*

9. Application: (4), Sea otters eat many of the organisms in the ecosystem.
10. Comprehension: (2), Darwin states that the low temperatures of the islands are probably due to the cold polar current.
11. Analysis: (2), The last red wolves were taken from the wild in 1980, and the first pups were born in captivity in 1988.
12. Evaluation: (2), The timeline shows that red wolf populations did not recover until both species protection laws and captive breeding programs were implemented.
13. Analysis: (5), Carnivores may eat herbivores, omnivores, or other carnivores.
14. Evaluation: (5), The paragraph describes differences between viruses and living things.
15. Analysis: (5), Unlike living things, viruses cannot reproduce themselves.
16. Evaluation: (2), The table shows that short-billed dowitchers eat marsh periwinkles, and marsh periwinkles eat cordgrass.
17. Application: (1), Marsh periwinkles rely on cordgrass for food, and short-billed dowitchers rely in part on marsh periwinkles for food.
18. Analysis: (2), The tracking devices can tell scientists whether the bear is still alive.
19. Application: (4), Water removes magnesium and calcium from the soil.
20. Analysis: (4), Food cannot enter the stomach unless it first passes through the esophagus.

UNIT 2 EARTH/SPACE SCIENCE

LESSON 1, *pp. 44–45*

1. Application: (3), If Wegener's hypothesis states that continents constantly move, in several million years they will be in different positions than they are today.
2. Analysis: (1), The "fit" of Africa and South America was one of the pieces of evidence Wegener used to support his hypothesis.
3. Application: (3), If the continents once had been joined, the same organisms would probably have lived on both continents. Finding fossils of the same organisms on both continents therefore would suggest that the continents once were joined.
4. Application: (1), Unlike a hypothesis, a theory explains a large number of observations.

LESSON 2, *pp. 46–47*

1. Comprehension: (5), The diagram shows that the crust is much thinner than any other layer of Earth.
2. Evaluation: (3), The diagram shows that melting happens in the mantle.
3. Comprehension: (1), The text and diagram indicate that volcanoes are common at plate boundaries.
4. Comprehension: (1), The text and diagram state that most erosion occurs where water flows fastest, and water flows fastest on the outside of a meander.
5. Application: (1), The diagram shows that rock and soil are deposited on the inside of a meander, and the text states that the water flows slowest on the inside of a meander.

LESSON 3, *pp. 48–49*

1. Comprehension: (4), The map shows the rivers in France.
2. Comprehension: (3), The map shows that New York City is located in the temperate forest biome.
3. Analysis: (5), Boreal forest is located north of temperate forest in the central and eastern parts of the United States.
4. Analysis: (2), The map shows that Seattle is in the mountain forest biome, and Los Angeles is in the chaparral biome. Different biomes have different climates.
5. Application: (3), Tundra and boreal forest are found in Alaska, which has a much colder climate than the rest of the United States.

LESSON 4, *pp. 50–51*

1. Comprehension: (3), The maps use different methods to show the elevation of land in the same area.
2. Application: (3), The maps show that Mount Saint Helens is located on the highest part of the region.
3. Analysis: (4), The topographic map shows that all three locations are between the 6,000-foot contour line and the 7,000-foot contour line.
4. Comprehension: (4), The maps show that Mount Saint Helens is located on the highest part of the region.
5. Application: (4), The maps show that the elevation on the north and south sides of the mountain is approximately 4,000 feet lower than the elevation at the summit.
6. Analysis: (3), The contour map shows that Mount Saint Helens has an elevation of about 8,000 feet, and the island has an elevation of about 6,000 feet.

LESSON 5, *pp. 52–53*

1. Comprehension: (3), The burial of plant matter by sediment is the second box in the flowchart.
2. Comprehension: (3), The flowchart shows that heat and pressure create metamorphic rock.
3. Analysis: (2), A cycle is, by definition, continuous.
4. Evaluation: (4), The flowchart shows that any rock type can be converted to any other rock type given the right conditions.
5. Comprehension: (2), The diagram shows that magma results from melting rock.
6. Application: (5), The diagram shows that heat and pressure produce metamorphic rocks.

LESSON 6, *pp. 54–55*

1. Analysis: (4), The stream is at the bottom of a canyon in the second diagram.
2. Application: (2), The two diagrams show that the spit became larger over time, so it is reasonable to assume that it did not exist prior to the first illustration.
3. Analysis: (1), Illustration B shows a baymouth bar in place of the spit in Illustration A. It also shows the longshore current flowing in the same direction in both illustrations. It's logical to assume that the current further built up the spit.

UNIT 2 *(continued)*

4. Evaluation: (4), The higher runoff in Stream A suggests that the soil in its watershed did not absorb as much rainwater.
5. Application: (5), The graphs show that there was more water in Stream A than in Stream B.

LESSON 7, *pp. 56–57*

1. Analysis: (2), A passage written by experts and supported by valid data is a reliable source of information.
2. Analysis: (4), The passage is based on scientific research, so it is most likely fact.
3. Application: (2), Some people may believe that a different theory is the best explanation for how the solar system was formed.
4. Comprehension: (2), The timeline shows that galaxies began to form around 12 billion years ago, not long after the Big Bang.
5. Application: (4), Data based on scientific research would provide the most accurate information.

LESSON 8, *pp. 58–59*

1. Comprehension: (4), Respected newspapers, such as the *New York Times,* are generally considered to be reliable sources of information.
2. Evaluation: (4), The most reliable source of information about a decision is the organization that made it.
3. Application: (2), Planetariums are generally good sources of scientific information about astronomy.
4. Analysis: (2), A source that gives mainly opinions is not a reliable source.
5. Analysis: (4), More recent information is generally more reliable than older information.

LESSON 9, *pp. 60–61*

1. Comprehension: (2), The paragraph describes problems caused by a lack of clean drinking water.
2. Analysis: (4), Improving sanitation will improve the health of people in sub-Saharan Africa.
3. Analysis: (5), The graph shows that only 1% of the water on Earth is fresh water.
4. Evaluation: (3), Conserving fresh water is the best way to make sure everyone has enough.
5. Analysis: (1), The paragraph states that fresh water is scarce in some parts of the world.
6. Application: (3), An assessment of the effectiveness of desalination must include an analysis of its costs and efficiency.

UNIT 2 REVIEW, *pp. 62–67*

1. Analysis: (2), Most sinkholes form when acidic water dissolves rock below the ground.
2. Evaluation (4), Limestone bedrock is most susceptible to chemical weathering by rainwater.
3. Evaluation: (1), Quotes from people involved in making a decision are generally good sources of information about the decision.
4. Application: (2), The person interviewed disagrees with the IAU's decision, and the article probably describes why.
5. Analysis: (2), The diagram shows vents on the side of the volcano, and lava comes out of vents.
6. Analysis: (2), The diagram shows that the volcano is made of layers of lava and ash.
7. Comprehension: (2), The diagram shows that very large hail is carried downward by a downdraft.
8. Analysis: (3), The diagram shows that both small and large pieces of hail can fall to the ground.
9. Analysis: (4), The diagram shows that hailstones rise and sink in clouds due to rising and sinking air.
10. Comprehension: (1), No evidence has been found for life on Mars and its presence there is speculative.
11. Application: (4), A diagram showing the locations and orbits of the planets would illustrate that the first group is closer to the sun and the second group is farther away.
12. Application: (2), For the test to be valid, all conditions except the condition being tested should be kept the same.
13. Evaluation: (1), The hypothesis that areas without trees and foliage are generally warmer than areas with trees and foliage can be tested and disproved.
14. Analysis: (2), The map shows that the landform has a lower elevation in the center than at the edges.
15. Analysis: (4), The contour lines on the western side of the landform are very close together, indicating that the western side of the landform is very steep.
16. Application: (2), The map shows that the landform is higher at the edges than in the middle.
17. Comprehension: (3), The map shows that Chicago is located in the grassland biome.
18. Comprehension: (1), The map shows that Boston is located in the temperate forest biome.
19. Analysis: (2), The chaparral and desert biomes are found in the warmest parts of the United States.
20. Analysis: (5), The map shows that the desert biome is found mainly in the western and southwestern areas.
21. Application: (1), If undrinkable water can safely be used to irrigate crops, then using drinkable water to irrigate them is a waste of drinkable water.
22. Evaluation: (2), The diagram shows that power plants are the second largest consumers of fresh water.
23. Application: (4), The diagram shows that the largest percentage of freshwater use in the United States is to water crops, including feed crops.
24. Evaluation: (5), The diagram shows that about 86% of the fresh water used in the United States is used for some form of industry or agriculture.

UNIT 3 PHYSICAL SCIENCE

LESSON 1, *pp. 70–71*

1. Analysis: (3), The diagram shows that the particles in solids are closer together than those in liquids or gases.
2. Analysis: (4), The diagram shows that the energy change between liquid and gas is much greater than the energy change between solid and liquid.

3. Analysis: (4), The diagram shows that approximately six times as much energy is needed to convert liquid to gas than is needed to convert solid to liquid.
4. Comprehension: (5), The diagram shows that the temperature of the system does not change during melting and evaporation.
5. Analysis: (2), A substance's melting and freezing points are the same. Point B represents the temperature at which water begins to melt.

LESSON 2, *pp. 72–73*

1. Analysis: (2), Elements with larger atomic numbers have greater atomic weights.
2. Comprehension: (5), Magnesium fluoride's boiling point is more than 800° C higher than the next highest boiling point.
3. Application: (5), The table shows that ionic compounds tend to have higher boiling points than covalent compounds.
4. Application: (2), Both hydrogen fluoride and hydrogen sulfide boil below room temperature, so they will exist as gases at room temperature.
5. Analysis: (5), All but two of the noble gases are used in lighting.
6. Evaluation: (2), The table's footnote indicates that some noble gases have specific uses in liquid form.

LESSON 3, *pp. 74–75*

1. Analysis: (4), Only the illustrated model shows how electrons are shared to form a hydrogen molecule.
2. Comprehension: (1), The paragraph states that the electron is located outside the proton and exists in a cloud around it.
3. Analysis: (2), Showing the atom as an illustrated model allows the location of the electron to be specified.
4. Application: (2), A three-dimensional model cannot be shown on two-dimensional paper.
5. Evaluation: (3), The left-hand model shows that the hydrogen atoms share two electrons to form a bond.
6. Application: (1), The ball-and-stick model can be used to show how the atoms in a molecule are positioned.

LESSON 4, *pp. 76–77*

1. Application: (3), If the water rose 1/5 of the way up the tube, the volume of gas in the tube must have decreased by 1/5; if only oxygen was removed from the air in the tube, then 1/5 of the air was oxygen.
2. Analysis: (4), The diagram shows that calcium reacted more strongly than zinc, and zinc reacted more strongly than copper.
3. Analysis: (3), Because both calcium and zinc reacted with acid, they must appear above hydrogen in the activity series.
4. Analysis: (2), When paper burns, new substances form.
5. Evaluation: (4), During a chemical reaction, new substances form.

LESSON 5, *pp. 78–79*

1. Analysis: (3), HBr, HI, HCl, and HF are all acids.
2. Application: (5), The first equation represents the mixing of two bases, which will probably produce no reaction. The second equation represents the mixing of an acid and a salt, which will probably produce no reaction. The third equation represents the mixing of an acid and a base, which will result in a reaction.
3. Analysis: (3), One molecule of HCl produces one hydrogen ion; one molecule of H_2SO_4 produces two hydrogen ions. Acidity increases as the number of hydrogen ions increases.
4. Application: (5), The Li and Cl ions will combine to form the salt. The hydrogen and hydroxide ions will combine to form water.
5. Application: (3), The reaction of an acid and a base produces a salt and water.
6. Analysis, (1): The K and Br ions must come from the base and the acid, respectively.

LESSON 6, *pp. 80–81*

1. Analysis: (3), The scooter's total displacement is 0 miles, because it returns to its starting point. It travels 8 miles total in 40 minutes, which is equivalent to 12 miles in one hour.
2. Analysis: (5), The product of 3 and 5 is 15.
3. Application: (3), If mass increases but force remains constant, acceleration must decrease.
4. Analysis: (4), The force the floor applies to the person is equal in magnitude to the person's weight.
5. Application: (4), A person with a mass of 65 kg has a weight of 637 newtons, and the upward force from the floor is equal to the person's weight.

LESSON 7, *pp. 82–83*

1. Analysis: (2), The diagram shows that the lever transforms a downward force into an upward force.
2. Analysis: (3), A single, fixed pulley does not change the amount of force required.
3. Application: (2), A block and tackle reduces the amount of force required to lift an object.
4. Analysis: (4), A pulley system that changes a downward force into an upward force would allow you to pull downward with your weight and use the resulting force to lift an object.
5. Analysis: (1), Simple machines do not change the amount of work that is done to lift an object.

LESSON 8, *pp. 84–85*

1. Analysis: (4), The graph shows that gas consumption in 2006 was slightly lower than it was in 2005.
2. Comprehension: (3), The two bars on the graph are the same length for the 2015 data point.
3. Analysis: (5), Showing two lines on the graph allows the two trends to be easily compared.
4. Analysis: (4), The graph shows that domestic oil production peaked in 1985 and has been decreasing ever since. It also shows that oil imports were at a minimum in 1985 and have been increasing ever since.

LESSON 9, *pp. 86–87*

1. Comprehension: (4), The diagram shows that the vast majority of U.S. energy comes from petroleum, natural gas, and coal.
2. Analysis: (4), According to the diagram, Russia produced approximately 1.5 quadrillion BTU of hydroelectric power in 2006, and China produced approximately 4.5 quadrillion BTU.
3. Analysis: (3), Each water droplet represents 1 quadrillion BTU, and Brazil's entry contains 3.5 water droplets.
4. Analysis: (2), Colorado and New Mexico show approximately the same fraction of a dime.
5. Analysis: (2), Each dime represents 10 cents per kilowatt-hour, and Utah's entry shows slightly more than half of a dime.

LESSON 10, *pp. 88–89*

1. Evaluation: (2), The crest is the distance between the resting position and the highest point on the wave.
2. Analysis: (5), As amplitude decreases, loudness decreases.
3. Analysis: (3), The amplitude of the wave is decreasing with time. As amplitude decreases, loudness decreases.
4. Evaluation: (3), Lawn mowers, jackhammers, and jet engines produce sounds above 85 dB.
5. Application: (1), Conversation has the lowest volume, so its sound waves must have the smallest amplitude.

LESSON 11, *pp. 90–91*

1. Comprehension: (2), If testing shows that a hypothesis is invalid, the hypothesis should be modified.
2. Analysis: (4), The table shows that organisms start to die at pH 6, and milk is the only substance with a pH higher than 6.
3. Application: (1), These data can be used to assess the effects of other acids, such as acid rain, on ecosystems.
4. Evaluation: (3), The data show that only four of the five liquids are acidic enough to be toxic.
5. Analysis: (1), The data show that the hypothesis is not completely valid, so it must be modified and retested.
6. Evaluation: (4), The data show that rainbow trout start to die at the highest (least acidic) pH.

LESSON 12, *pp. 92–93*

1. Analysis: (4), The diagram shows that most of the iron filings are concentrated at the ends of the magnet.
2. Analysis: (2), The diagram shows that a north pole attracts a south pole but repels another north pole.
3. Comprehension: (5), The paragraph states that magnetic field lines run in a specific direction.
4. Analysis: (4), If like poles are together, the magnets will repel each other; if opposite poles are together, the magnets will attract each other.
5. Application: (3), Iron filings will align themselves with a magnet's magnetic field.

UNIT 3 REVIEW, *pp. 94–99*

1. Analysis: (2), The mechanical advantage of Machine B is 10; this is the largest mechanical advantage of the machines listed.
2. Analysis: (2), The product of 25 and 4 is 100.
3. Analysis: (3), If the rays are impossible to detect, there is no way to show that they do or do not exist.
4. Application: (1), Scientific theories are rejected in favor of new theories if the new theories can explain new data that the older theories could not.
5. Application: (2), The leftward force is much greater than the rightward force.
6. Analysis: (4), Potential energy decreases as the point on the ramp decreases.
7. Evaluation: (1), The graph shows that potential energy is highest when kinetic energy is lowest.
8. Application: (3), If the direction of the current is reversed, the magnetic field polarity also will reverse.
9. Analysis: (5), The particles in a transverse wave vibrate perpendicular to the wave's direction of travel.
10. Evaluation: (1), Longitudinal waves cause particles to vibrate parallel to the wave's direction of motion.
11. Application: (2), Molecule B contains three carbon atoms.
12. Analysis: (5), Each full circle represents 10% of energy consumption. Color TVs have 3.5 circles, and TV accessories have 1.5 circles.
13. Analysis: (1), In a single-displacement reaction, a pure element reacts with a compound to form a new compound and a different pure element.
14. Comprehension: (4), In a double-displacement reaction, the ions in two compounds switch places.
15. Analysis: (5), Point D is located in the region of the graph that represents the vaporization of liquid water into gas.
16. Analysis: (5), Of the answer choices, vinegar has the lowest pH.
17. Analysis: (2), Of the answer choices, ammonia has the highest pH.
18. Application: (3), NaOH is a base, and H_2SO_4 is an acid.
19. Application: (1), The solution must contain a base, and magnesium hydroxide is a base.

Index

Note: Page numbers in **boldface** indicate definitions or main discussion. Page numbers in *italic* indicate a visual representation. Page ranges indicate examples and practice.

A

Acceleration, *81*
Acids
  characteristics of, *78–79*
  effect on living things, *91*
  pH values, *91, 99*
  reactions with metals, *77*
Adaptations, *17*, 28, 37
Alcohols, *97*
Algal blooms, 14
Alkali metals, *72*
Amphibians, *3*
Amplitude, *89*
Analyzing
  information, **16**, 16–17
  results, **90**, 90–91
Animalia, *19*
Animals
  adaptations, *17*, 28, *29*
  biomes of, *17*
  black bears' diet, *4*
  body systems, *9*
  carnivores/herbivores/omnivores, *39*
  categorizing/classifying, *3*, *18,–21*
  cells of, *9*
  characteristics of major groups of, *3*
  chordates, *2–3*
  classes of mammals, *20–21*
  as consumers, 32
  dependence on plants, 12
  diet of, *4, 7, 40*
  ecosystems of, *6–7*, 32–33, *38, 40*
  endangered species, *5, 24–25, 39*, 41
  extinction, *25*
  feeding systems, *4, 6–7*, 32, *38–39*
  kingdom of, *18*
  pollution and, 14–15
  populations of, *5*, 33, *36, 38*
  relationships among, *34–35*
  tracking of, 41
  See also Organisms; *specific animal*
Annelida, *2*
Antibodies, 27
Applying concepts, **92**, 92–93
Aquatic ecosystems, 14
Archaebacteria, *19*
Arthropoda, *2*
Asthenosphere, *47*
Atmosphere
  greenhouse gases, *85*
  hail formation, *64*
  pollution in, 15
Atomic number, *73*
Atoms, *74–75*
Attraction, *93*
Axis labels, 84

B

Bacteria
  beneficial activities of, 10, *11*
  cells parts of, *8*
  disease caused by, 27
  infection and, *26*
  kingdom of, *19*
  shapes of, *11*
Bald Eagle Protection Act, *24, 25*
Bar graph, 5
Bases, *78–79, 99*
Bears, *4*, 41
Beneficial bacteria, 10, *11*
Big Bang, *56–57*
Biomes, *16–17, 49, 66*
Bipolarity, *93*
Birds, *2, 3, 29*
Black bears, *4*
Block-and-tackle pulley, *83*
Body systems
  circulatory system, 9
  digestive, 10, *22–23, 41*
  immune, *26, 27*
  nervous, 9
Boiling point, *73*

C

Cactus, 13
Calculations, interpreting outcomes with, **80**, 80–81
Canyon formation, *54*
Carbon, 97
Carbon dioxide, *85*
Carnivora, *20*
Carnivores, *39*
Categorizing, **18**, 18–19
Cause and effect, **26**, 26–27
Cells, *8–9*
Charts, **4**, *4–5*
  flowcharts, **52**, 52–53
Chemical reactions, *76–77, 98*
Chiroptera, *20*
Chordates, *2–3*
Circle graphs, 4
Circulatory system, 9
Classes of organisms, *19*
Classifying, **18**, 18–19
Clean Air Act, 15
Climate
  of biomes, *16–17*
  of ecosystems, 33
  effect on living things, *5*
  of Galápagos Islands, 38
  greenhouse gases and, *85*
Cnidaria, *2*
Coal formation, *52*
Coastlines, *55*, 56
Commensalism, 35
Communities, 33
Comparing, **20**, 20–21, **54**, 54–55
Compass rose, *48*
Composite volcanoes, *63*
Compounds
  covalent, 73
  ionic, 73
  organic, 97
Concepts, applying, **92**, 92–93
Conclusions, drawing, **30**, 30–31, 32
Condensation, *71*
Consumers, 32
Context clues, **14**, 14–15
Continental drift, 44, *45*
Contour lines, *50, 65*
Contrasting, **20**, 20–21, **54**, 54–55
Core of Earth, *46*
Covalent bonds, 73, *75*
Covalent compounds, 73
Cowpox, *27*
Crust of Earth, *46–47*
Current, electric, 96
Cutaway diagrams, 8, **46**, 46–47

D

Darwin, Charles, 28–29, 38
Data
  looking for patterns/trends in, 78
  range of, 84
  See also Graphs; Tables
Decibels, *89*
Deciduous forests, *16*
Decomposers, 10
Deposition, *54–55*
Desalination, 61
Desert biome, *16, 49*
Desert ecosystem, *7, 36*, 37
Details, **10**, 10–11, 12–13
Diagrams
  categorizing/classifying organisms, *19*
  demonstrating concepts, 92
  interpreting, **6**, 6–7, **70**, 70–71
  with numerical values, *80*
  relating text to, 88
  scale of, 46
  of sequences of events, 22
  three-dimensional, **46**, 46–47
Diet of animals, *4, 6–7*, 39, *40*
Digestion, 10, *22–23, 41*
Diseases, 10, *26–27*, 40
Displacement, *80*
Distance, speed and, *80*
Dominant traits, *37*
Drawing conclusions, **30**, 30–31, **82**, 82–83
Dwarf planets, 58, 59

INDEX

E

Eagles, *24, 25*
Earth
biomes of, *16–17, 49, 66*
continental drift, 44, *45*
crust of, *46–47*
earthquakes, 29, *47*
erosion of, *47, 54*, 56
layers of, *46–47*
nitrogen cycle, *11*
plate tectonics, 45, *47*
rock cycle, *53*
volcanoes, 29, 45, *47, 51, 63*, 64
Earthquakes, 29, *47*
Earth/Space Science
relevance of, 43
See also specific topic in unit two
Ecosystems
animal feeding roles in, *39*
freshwater, *40*
grasslands, *6, 36*
kelp forest, *38*
organisms of, 32–33
pollution of, 14
role of plants in, 12
wetlands/riparian areas, *15*
Effect, cause and, **26**, 26–27
Electricity
consumption of, *98*
magnetism and, *96*
prices of, *87*
Electron microscope, 9
Electrons, *74–75*
Elements
atoms of, *74–75*
Hydrogen, *75*
Noble gases, *73*
Endangered species, *5, 24–25, 39*
Endangered Species Act, *24, 25*
Endangered Species Conservation Act, *24*
Endangered Species Preservation Act, *24*
Energy
change of state and, *71*
consumption of, *84–86*
electricity, *87, 98*
heating curve of water, *71*
movement through ecosystems, 32
potential/kinetic, *96*
price of electricity, *87*
sources of, *86–96*
Energy cycle, *32*
Environmental damage, 14
Equal and opposite forces, *81*
Erosion, *47, 54*, 56
Eubacteria, *19*
Evaluating
hypotheses, **44**, 44–45
information, **58**, 58–59
Evaporation, *71*
Evolution, **28**, *29*

F

Facts
generalizations vs., 32
opinion vs., **56**, 56–57
Families of organisms, *19*
Feeding systems, *6–7*
Figures. *See* Visuals
Finches, *29*
Fish, *2, 3*
Floods, 56
Flowcharts, **52**, 52–53, 60
Food chain, *6–7*, 12
Food web, *7*
Footnotes, 72
Force, *81*
input/output, *82*
of magnetic fields, *92*
Newton's second law of motion, *81, 95*
Newton's third law of motion, *81*
output/input, *94*
simple machines and, *82–83, 94*
Fossil fuels, 64, *85*
Freezing, *71*
Freshwater ecosystem, *40*
Frontmatter, *40*
About the GED Tests, iv–v
subject-area breakdown, iv
About GED *Xcelerator*, vi
About *Xcelerator Science*, vii
Before You Begin: Using Logic and Making Assumptions, x–xi
Copyright/Acknowledgements, ii
Study Skills, ix
Table of Contents, iii
Test-Taking Tips, viii
Title Page, i
Fulcrum, *82*
Fungi, *19*

G

Galápagos Islands, 29, 38
Gases, *70*
GED Journeys
Carmona, Richard H., BLIND
Musgrave, Story F., 42
Patrick, Danica, 68
GED Tests
construction of, iv
number of people taking/year, vi
number of people who passed in 2007, vi
preparation for, ix
skills required for, x
subjects tested/number of questions per subject, iv
time limits for, iv
Generalizing, **32**, 32–33
Genetics, *30–31*, 37
Genus, *19*
Germ theory, 27
Graphics. *See* Visuals
Graphs
interpreting, **4**, 4–5
multi-bar, **84**, 84–85
multi-line, **84**, 84–85
pictographs, **86**, 86–87
Grasslands, *6, 7, 16, 36, 49*
Greenhouse gases, *85*
Grey wolves, *5*

H

Habitat, *5*
Hail formation, *64*
Hash marks, 24
Headings of diagrams, 70
Hearing damage, 89
Herbivores, *39*
Heredity, *30–31, 37*
Hibernation, 17
Histamine, *26*
Humans, *2*. *See also* Body systems
Hydrogen, *74–75*
Hydronium ions, *99*
Hydroxide ions, *78, 79, 99*
Hypotheses, **44**, 44–45, 95
formation of, 78
testing of, 90, 91

I

Identifying problem and solution, **60**, 60–61
Igneous rock, *53*
Illustrated models, **74**, 74–75, 92
Illustrations, **8**, 8–9
Immune system, *26*, 27
Indicators, *99*
Infection, *26*
Inferences, making, **28**, 28–30
conclusions based on, 82
hypotheses formulation and, 44
Inflammatory response, 26
Information
analysis of, **16**, 16–17
evaluation of, **58**, 58–59
left out of texts, 82
on maps, 48
Inherited traits, *30–31, 37*
Input force, *82, 94*
Interferons, 27
International Astronomical Union (IAU) Resolution, 59, 63
Interpreting
charts and graphs, **4**, 4–5
complex diagrams, **70**, 70–71
complex tables, **72**, 72–73
diagrams, **6**, 6–7
flowcharts, **52**, 52–53
illustrations, **8**, 8–9
multi-line graphs, **84**, 84–85
multi-bar graphs, **84**, 84–85
observations, **76**, 76–77
physical/topographical maps, **50**, 50–51
pictographs, **86**, 86–87

tables, **2**, 2–3
three-dimensional diagrams, **46**, 46–47
timelines, **24**, 24–25
using calculations to interpret outcomes, **80**, 80–81
using established concepts, 92
Intervals on timelines, 24
Ionic compounds, 73
Ionization reactions, *78–79*
Iron oxide, *76*

J

Jellies, *2*
Jenner, Edward, 27

K

Kelp forest, *38*
Key
for bar graphs, 84–85
for circle graphs, 4
for complex tables, *72*
for line graphs, 85
for maps, *48*
for models, 74–75
for pictographs, *86*
Kilograms, *81*
Kinetic energy, *96*
Kingdoms of organisms, *18–19*

L

Labels
on circle graphs, 4
on diagrams, 46, 88
on graphs, 84
of illustrations, 8
Lacey Act, *24*
Landforms
formation of, 45, *47*
maps of, *50–51*
Laws, *24–25*
Leader lines, 46
Lever, *82*
Life Science
relevance of, 1
See also specific topic in unit one
Light microscope, 9
Line graph, 5
interpreting multi-line, **84**, 84–85
Linnaeus, Carolus, 19
Liquids, *70*
Lithosphere, *47*
Litmus paper, *99*
Living things, *6*, 32–33. *See also* Animals; Plants
Longitudinal waves, *89, 97*
Longshore currents, *55*
Lysosomes, *9*

M

Magnetic fields, *92–93, 96*
Magnetism, *92–93, 96*
Magnets, *92–93,* 96
Main idea
of complex diagrams, 70
identifying, **10**, 10–11
summarizing, **12**, 12–13
Making assumptions
about chemical reactions, 76
about information from multiple sources, 82
about information on graphs, 84
about lines on maps, 48
about scale diagrams, 46
about scale of illustrated models, 74
applying concepts, 92
changes in rivers, 54
determining fact from opinion, 56
relating text and figures, 88
units in complex table, 72
using timelines, 24
Making generalizations, **32**, 32–33, 78
Making inferences, **28**, 28–29
Mammals
as chordates, *3*
classes of, *20–21*
endangered species, *5*
Mantle of Earth, *46*
Maps
definition of**, 48**
map symbols, *48–49*
physical maps, **50**, 51–52, 66
topographic maps, **50**, 51–52, 65
Marsupials, *21*
Mass, *81*
Matter
change of state, *71*
states of, *70*
waves and, *97*
Mechanical advantage, *94*
Mechanical waves, *89*
Melting, *71*
Mendel, Gregor, 30–31
Metals, *76–77*
Metamorphic rock, *53*
Microbes, *26*, 27, 40. *See also* Bacteria; Viruses
Migrating Bird Treaty Act, *24, 25*
Migration, 17
Mitochondria, *9*
Models, **74**, 74–75
Mollusca, *2*
Monotremes, *21*
Mountain formation, 45, 47
Mount Rainier, *50*
Mount Saint Helens, *51*
Movable pulley, *83*
Multi-bar graphs, **84**, 84–85
Multi-line graph, **84**, 84–85
Mutualism, *34*

N

Nervous system, 9
Neurons, *9*
Newtons, *81*
Newton's laws of motion, *81, 95*
Nitrogen cycle, *11, 37*
Nitrogen-fixing bacteria, *11*
Noble gases, *73*
Nonliving things, *6,* 32–33
Nonvascular plants, 13
Nucleus, *9*
Numbering of events in sequence, 22

O

Observations, interpreting, **76**, 76–77
Oil consumption, *85*
Omnivores, *39*
Opinion, fact vs., **56**, 56–57
Order of organisms, *19*
Organisms
adaptations, *17,* 28, *29,* 37
categorizing/classifying, *2–3, 18–21,* 33
Darwin's theory concerning, 28, *29*
ecosystems of, 12, 14–15, 32–33, 40
in fresh water ecosystems, 40
heredity, *30–31, 37*
relationships among, *34–35, 38*
See also Animals; Plants
Origin of species, 28, *29*
Outcomes
predicting, **78**, 78–79
using calculations to interpret, **80**, 80–81
Output force, *82, 94*
Oxygen, 12

P

Pangaea, 44
Particles, *70*
Pasteur, Louis, 27
Percentages on circle graphs, 4
Phloem, 13
Photosynthesis, 12
pH values, *91, 99*
Phyla, *19*
Physical changes, *77*
Physical maps, **50**, 50–51
Physical Science
relevance of, 69
See also specific topic in unit three
Pictographs, **86**, 86–87
Pitch, 89
Placental mammals, *21*
Planets, 58–59, 63
Plantae, *19*
Plants
adaptations of, 17, 37
biomes of, *17*
cactus, 13

INDEX

categorizing/classifying, *13, 18–19*
ecosystems of, *6*, 32–33, *40*
endangered species, *5*
environments of, *12*
feeding systems, *6*
in food chain, 12
in nitrogen cycle, *11*
nitrogen cycle and, *11*
photosynthesis, *12*
pollution and, 14, *15*
as producers, 32
relationships among, 35
role in survival of organisms, 12
vascular/nonvascular, *13*
See also Organisms
Plate tectonics, 45, *47*
Pluto, 58–59, 63
Polar bear, 41
Polarity, *93*
Poles of magnets, *92–93*
Pollution, 10, 14–15
Populations, *5*, 33, *35, 36, 38*
Porifera, *2*
Potential energy, *96*
Precipitation, *16, 17*, 33, *55*
Predator-prey relationship, *35*
Predicting outcomes, **78**, 78–79
Problem and solution, **60**, 60–61
Producers, 32
Products, 76
Proteins, 27
Protisa, *19*
Protons, *74–75*
Pulley, *83*
Punnett square, *31, 37*

Q

Questioning, **34**, 34–35

R

Radio collars, 41
Rainfall. *See* Precipitation
Range of data, 84
Reactants, 76
Reading skills
analyzing of information, **16**, 16–17
categorizing, **18**, 18–19
cause and effect, **26**, 26–27
classifying, **18**, 18–19
comparing and contrasting, **20**, 20–21
determining fact from opinion, **56**, 56–57
drawing conclusions, **30**, 30–31
identifying problem/solution, **60**, 60–61
information, analysis of, **16**, 16–17
information, evaluation of, **58**, 58–59
interpreting timelines, **24**, 24–25
main idea and details, **10**, 10–11
making generalizations, **32**, 32–33
making inferences, **28**, 28–29
questioning, **34**, 34–35
sequences, **22**, 22–23
summarizing, **12**, 12–13
understanding/evaluating hypotheses, **44**, 44–45
using context clues, **14**, 14–15
Recessive traits, *37*
Red wolf conservation, *39*
Relating text and figures, **88**, 88–89
Reptiles, *3*
Repulsion, *93*
Results, **90**, 90–91
Rock cycle, *53*
Rows
in complex tables, 72
in tables, 2
Runoff, 15, *55*
Rust, *76*

S

Salivary glands, 23
Salts, *78*
Scale bar, *48*
Scale diagrams, 46
Scientific method, 78, 90–91
analyzing results, **90**, 90–91
applying concepts, **92**, 92–93
interpreting observations, **76**, 76–77
interpreting outcomes, **80**, 80–81
predicting outcomes, **78**, 78–79
questioning, **34**, 34–35
understanding/evaluating hypotheses, **44**, 44–45
Sedimentary rock, *53*
Sequences, **22**, 22–25
Simple machines, *94*
Single, fixed pulley, *83*
Sinkholes, *62*
Smallpox, *27*
Solar system, 56–57, 59
Solids, *70*
Solutions, problems and, **60**, 60–61
Sound, *89*
Sources of information
drawing conclusions from many, **82**, 82–83
evaluating, 58, 63
Species, *19*
Speed, *80*
Sponges, *2*
States of matter, *70–71*
Substances, chemical reactions between, *76–77*
Summarizing, **12**, 12–13, 30
Sunlight, 12, 33
Supporting details, **10**, 10–11, **12**, 12–13
Symbiosis, 35
Symbols
in complex tables, 72
of elements, *73*
on pictographs, 86

T

Tables, **2**, 2–3
interpreting complex ones, **72**, 72–73
using to categorize/classify, *18–19*
Tectonic plates, 45, *47*
Temperatures
of biomes, *16, 17*
of ecosystems, 33
in Greenland, *57*
in rural vs. urban areas, 65
Test-taking tips
categorizing/classifying organisms, 18
converting units, 80
evaluating sources, 58
first step of sequences, 22
flowcharts of problems/solutions, 60
interpreting diagrams, 6
interpreting graphs, 4
interpreting illustrations, 8
interpreting pictographs, 86
making generalizations, 32
read all answer choices, 10, 36, 62, 94
read each question carefully, 36, 62, 94
read titles of tables, 2
using a map key, 50
validity of conclusions, 30
words used for comparing/contrasting, 20
Text, relating to figures, **88**, 88–89, 92
Theory, **44**, 95
Three-dimensional diagrams, **46**, 46–47
Tidal waves, 29
Time, speed and, *80*
Timelines, **24**, 24–25, *39*
Titles
of complex diagrams, 70
of tables, 2
Tongue, 23
Topographical maps, **50**, 50–51, *65*
Transverse waves, *88–89, 97*

U

Unbalanced forces, *81, 95*
Understanding
hypotheses, **44**, 44–45
illustrated models, **74**, 74–75
maps/map symbols, **48**, 48–49
Unit reviews, 36–41, 62–67, 94–99
Units of measure, *81*, 89
Using logic
cause-and-effect relationship, 26
combining text/graphic information, 16
identifying hypotheses, 44
inferring relationships in diagrams, 70
interpreting *bloom* from context, 14
looking for patterns/trends in data, 78
recognizing inferences, 28
scientific method, 90
understanding flowcharts, 52
using/recognizing synonyms, 12

V

Vaccinations, 27
Vascular plants, *13*
Viruses, 27, 40
Visuals
analysis of information in, 16–17
charts and graphs, **4**, 4–5
comparing and contrasting, **54**, 54–55
complex diagrams, **70**, 70–71
complex tables, **72**, 72–73
diagrams, **6**, 6–7
drawing conclusions from, 82
illustrated models, **74**, 74–75
illustrations, **8**, 8–9
maps, 48–51, **48**, **50**
multi-bar graphs, **84**, 84–85
multi-line graph, **84**, 84–85
pictographs, **86**, 86–87
questioning, 34
relating to text, **88**, 88–89, 92
tables, **2**, 2–3
three-dimensional diagrams, **46**, 46–47
Volcanoes, 29, 45, *47*, *51*, *63*
Volume of sound, *89*

W

Water
from acid and base combination, *99*
desalination, 61
erosion by, *47*, *54*
fresh water supplies, 60, *61*, *67*
heating curve of, *71*, *98*
as need of living things, 33
plants and, *13*
pollution of, 14, *15*
as product of ionization reactions, *79*
quality improvement, *15*
runoff, 15, *55*
Wavelength, *89*
Waves, *88–89*, *97*
Wegener, Alfred, 44–45
Weight, *81*
Wetlands, *15*
Wildlife refuge, *24*

X

***X*-axis**, 84
Xylem, 13

Y

***Y*-axis**, 84

INDEX